BLOCK POLYMERS

BLOCK POLYMERS

Proceedings of the Symposium on Block Polymers
at the Meeting of the American Chemical Society
in New York City in September 1969

Edited by S. L. Aggarwal

Manager, Materials Research and Technical Services
The General Tire and Rubber Company
Research and Development Center, Akron, Ohio

Ⴔ PLENUM PRESS · NEW YORK-LONDON · 1970

CHEMISTRY

Library of Congress Catalog Card Number 74-119054
SBN 306-30481-3

© 1970 Plenum Press, New York
A Division of Plenum Publishing Corporation
227 West 17th Street, New York, N. Y. 10011

United Kingdom edition published by Plenum Press, London
A Division of Plenum Publishing Company, Ltd.
Donington House, 30 Norfolk Street; London W.C. 2, England

PREFACE

Block polymers represent another milestone in the preparation
of polymers of controlled structure. Catalysts and polymerization
methods that allowed the preparation of polymers in which the
stereo- and geometric isomerism of the monomer units could be controlled have indeed been among the major developments in polymer
science during the last decade. The synthesis of block polymers,
in which the sequence length of the comonomer units can be controlled, portends equally important developments in the science and
technology of polymers.

The papers collected in this volume cover primarily the proceedings of the most recent symposium on block polymers, sponsored
by the Division of Polymer Chemistry of the American Chemical
Society. It was held in New York City during the Society's 158th
National Meeting in September, 1969. Additional contributions from
selected authors were invited especially for this book to achieve
the most up-to-date account of the advances that have been made
since the development of the thermoplastic elastomers that first
brought into focus this important area of research.

The first two papers in this volume draw attention to the
various problems that should be considered in the preparation of
block polymers of precisely defined structure from styrene and
butadiene or isoprene by anionic polymerization. Characterization
of block polymers presents many problems and there is a paucity of
systematic work in this area. Attention has been given to the dilute solution properties of block polymers,however, in one of the
papers in this volume.

From transition behavior and direct electron microscopical
studies, the multiphase structure in the solid state has been established as one of the common features of a variety of block polymers.
An important advance has been the development of a theoretical frame-
work which provides some insight into the factors that control the
morphological details of the domains originating from phase separation in block polymers. These studies have been extended to investigate the use of simple diblock copolymers in the blending of

v

two different homopolymers. This behavior is analogous to the well
known use of surfactants in colloidal dispersions. These aspects
of block polymers are the subjects of several papers in this volume.

Until recently, most of the interest in block polymers had been
confined to those of styrene and butadiene or isoprene. A number of
papers in this volume report on new block polymer systems. Noteworthy among these are block polymers based on ethylene oxide-
styrene, siloxanes, propylene sulfide, imino ethers, and α-methyl
styrene-isoprene. The block structure of some of the polyurethanes
is discussed in two of the papers. Another paper describes the
use of cationic catalysts in the preparation of block polymers
based on tetrahydrofuran, and discusses the kinetics of polymerization of tetrahydrofuran.

The editor had hoped to include some papers on the applications
of block polymers, and regrets that no papers were offered dealing
specifically with this subject. It is hoped that this volume will
stimulate future systematic studies on the applications of block
polymers.

The expeditious publication of this volume soon after the symposium has been possible only because of the special photo-offset
method of printing from the original manuscripts of the authors.
Because of this, the editor had to be content with a number of
editorial compromises. One of these is the lack of uniformity in
the use of the terms, "block polymers" and "block copolymers". It
is hoped that in the context of the various papers, the inter-
changeable use of these terms will not distract the reader or result in any ambiguity.

I am grateful to all the authors and the publishers for their
cooperation in expediting the publication of this volume. If this
publication stimulates further research in block polymers, the ef-
forts of all will be amply rewarded.

S. L. Aggarwal
Akron, Ohio

February 24, 1970

CONTENTS

PROBLEMS IN THE SYNTHESIS OF BLOCK

POLYMERS BY ANIONIC MECHANISM

Maurice Morton

Institute of Polymer Science

The University of Akron

The recently developed "thermoplastic elastomers"
owe their properties to their block polymer structure.
As is well known, these can be defined as SDS block polymers, where S represents a polystyrene block while D
represents a polydiene block (generally butadiene or
isoprene). The result is that these linear polymers
show similar behavior to that of crosslinked networks,
presumably due to the fact that the polystyrene end
blocks aggregate into "domains", because of their incompatibility with the polydiene, and these glassy domains
can act as "network" junctions at ambient temperatures.

Recent studies[1] have shown that this is indeed the
morphology which is found in these systems. Furthermore,
the physical mechanisms[2] which operate in these systems
and the effect of the molecular architecture[3,4] on these
mechanisms has now been considerably elucidated. It
appears that, in order to attain the optimum elastomer
properties, the block lengths are fairly well prescribed,
the polystyrene being in the range of 10,000-15,000 in
molecular weight, while the polydiene should be in the
50,000-100,000 range. Furthermore, the elastic properties
and tensile strength of these materials appears to depend
on the ability of the polystyrene domains to absorb the
applied stresses, undergoing an inelastic deformation in
the process. Hence it is obvious that these properties
will depend greatly on this domain formation, which is,
of course, a function of the incompatibility of the two
types of blocks. The latter, in turn, depends not only

1

on the chemical nature of the blocks but on their size,
especially in the ranges indicated, e.g., polystyrene
blocks below 10,000 in molecular weight may result in a
drastic decrease in tensile strength of the material.
It is therefore important to realize that such factors
as block size distribution and efficiency of block polymer formation can have a profound effect on the final
properties of these materials. The polymerization system
used for the synthesis of these polymers must therefore
meet certain specifications.

The synthesis of block polymers, such as SDS, having
a precise and predictable structure, only became possible
with the advent of homogeneous anionic polymerization of
the non-terminating type.[5] Such "living" polymers have
made it possible to use four different approaches to the
synthesis of SDS polymers, based on type of initiator
used:

1. Difunctional initiators, e.g., sodium naphthalene,
    leading to a "two-stage" process (i.e., polymeriza-
    tion of D followed by polymerization of S).
2. Three-stage process, using monofunctional initiators,
    e.g., alkyllithium.
3. Two-stage process, using monofunctional initiators
    to synthesize SD diblocks and subsequent coupling
    to SDS.
4. Two-stage process, with alkyllithium initiators,
    involving formation of an initial S block followed
    by copolymerization of the styrene and diene, in
    which the latter is preferentially polymerized.

Each of the above methods has advantages and dis-
advantages which have a direct bearing on the quality
of the final polymer. These will be considered, with
special reference to the SDS polymers, in connection
with the following parameters: a) initiation and
termination problems, b) microstructure of polydiene.

1. Difunctional Initiators
    These initiators, e.g., sodium naphthalene, dilithium
    compounds, etc., are, in principle, best for the
    synthesis of ABA block polymers, on two counts. In
    the first place, they involve only a two-stage
    process, i.e., the sequential addition of two
    monomers, B followed by A, thus minimizing any
    termination of blocks by adventitious impurities
    present in the monomers. It should be noted that
    termination of a B block at one end only will lead

to an AB polymer while the statistical termination
of some of the B blocks at both ends will lead only
to B homopolymer. It will be seen later that
presence of the latter does not have a profound
effect on the behavior of the ABA polymer while the
AB diblock polymer does have a very great effect.

Another marked advantage of this system is in the
case of "unidirectional" block polymerization, i.e.,
where B can initiate A but not vice versa. Thus,
for example, it is possible in this way to synthesize
a block polymer where B is a vinyl or diene block
while A is an ethylene oxide block.

The main problem in the use of such difunctional
initiators for SDS block polymers is due to the
fact that they are generally soluble only in ethers,
or similar solvents, which lead to a low 1,4 structure in the polydiene, i.e., a non-rubbery polymer.
It is possible, of course, to use metallic lithium
itself as an initiator, in hydrocarbon solvents, and
thus achieve a high 1,4 structure, but the extremely
slow initiation rate of such heterogeneous systems
makes it impossible to control the chain length
within the prescribed limits for these block polymers.

In this connection, some new developments[4] in homogeneous polymerization with dilithium initiators
have made it possible to circumvent these difficulties, and to use this system to generate ABA
block polymers of high purity. The soluble dilithium initiators were prepared in high concentration
by the reaction of lithium with 1,1-diphenylethylene
in presence of a small proportion of aromatic ether,
and these were used to synthesize an ABA polymer
from isoprene (B) and α-methylstyrene (A). This
method is particularly important in the case of a
monomer like α-methylstyrene, which has a low
ceiling temperature and therefore requires low
temperatures to avoid the presence of residual
monomer. In this particular case, the isoprene was
first polymerized by means of the dilithium initiator, leading to a high 1,4 structure (∿90%), after
which a substantial amount of dimethoxyethane was
added to permit the polymerization of the
α-methylstyrene to virtual completion at -78°C.

2. Three-stage Process with Monofunctional Initiators
This is typified by the organolithium polymerization
systems, which can be used in a variety of solvents.

They are suitable only for "reversible" block
polymerizations, i.e., where both A and B blocks
can initiate each other. In the case of the SDS
systems, they can be used in hydrocarbon solvents to
achieve high 1,4 structures in the D block. They have
the disadvantage of involving three sequential monomer
additions with the attendant dangers of termination
by impurities. Furthermore, attention must be paid
to the relative rates of initiation and propagation
of each block. Actually, it is the initiation of
the first block, S, that poses the only problem,
since this is a short block and requires a very fast
initiation rate in order to assure a narrow molecular
weight distribution. Such fast initiation can be
accomplished either by the use of sec-butyllithium
or by means of primary alkyllithiums together with
small proportions of an aromatic ether, like anisole, which markedly accelerates the initiation rate
without disrupting the high 1,4 chain structure.[3]
The subsequent initiation of the diene block by the
styryl lithium is very rapid,[6] too, either in
presence or absence of the aromatic ether, while the
initiation of the last styrene block can be made very
fast in presence of any added ether, either aromatic
or aliphatic.

The efficacy of such a three-stage process in producing a "pure" block polymer is indicated by the
Gel Permeation Chromatographs shown in Figure I for
SBS and SIS polymers[3] initiated by ethyl lithium in
benzene containing a small proportion of anisole.
The sharpness of the single peaks, the absence of
any shoulders, and the fact that the $\overline{M}_n$ of these
polymers was found to be within 5% of the stoichiometric value attests to the absence of any fortui-
tous termination under the rigorous high-vacuum
purification techniques used.

3.   Monofunctional Initiation and Coupling[7]
     This is a two-stage process, analogous to that of
difunctional initiation and, similarly, is useful
for "unidirectional" block polymerization where, for
example, A can initiate B but not vice versa. It
also has the advantage of a two-stage process in
minimizing termination. However, it has one very
great disadvantage in that it demands a high precis-
ion in the stoichiometry of the coupling reaction
and in its efficiency. Furthermore, any deficiency
in the coupling reaction can lead to the formation

of excess diblocks (SD) which have been shown[8] to be very deleterious in very small amounts.

Figure 2 shows a GPC trace of a vinyl biphenyl-isoprene-vinyl biphenyl polymer[9] prepared by coupling a 4-vinyl-biphenyl-isoprene lithium diblock by means of phosgene. The presence of a small proportion (∼5%) of poly-4-vinyl-biphenyl as well as a substantial amount (26%) of the uncoupled diblock is obvious.

4. Two-stage Process of Block Polymerization and Copolymerization of Styrene and Dienes by Organolithium

This method is based on the known behavior of styrene in copolymerization with dienes in lithium-hydrocarbon systems,[10] i.e., the dienes are preferentially polymerized first so that most of the styrene ends up as a homopolymer block at the end. It has the usual advantage of a two-stage system in minimizing the effects of termination. Thus, even if the monomers contain terminating impurities, these may not be particularly deleterious. The first addition of styrene may lead to destruction of some of the organolithium initiator but this will merely increase the polystyrene block length to some extent. The second addition, of the proper mixture of styrene and the diene, may result in termination of the first styrene block, i.e., formation of free polystyrene, which is known[8] not to be too important in affecting the polymer properties. Thus no free diblocks should form, if the termination reaction occurs rapidly relative to propagation.

The marked disadvantage of this method is due to the fact that some proportion of styrene becomes copolymerized with the diene, and this decreases the incompatibility of the two phases, leading to more phase blending and hence poor tensile properties. This is especially noteworthy in SIS polymers, since isoprene does not exclude the styrene as strongly as does butadiene. Hence the "isoprene block" contains a substantial amount of styrene, and this is reflected in the poorer physical properties of these polymers.[11]

Effect of Incomplete Block Polymerization

In considering which of the above methods would be most suitable for use, it would be advantageous to have

some idea of the relative importance of the presence of
polymeric "impurities" in these block polymers. Some
recent results[8] have yielded such information with
regard to the effect of the presence of free S blocks,
free I blocks and free SI diblocks in an SIS system.
These results are shown in Tables I and II.

TABLE I

EFFECT OF ADDED HOMOPOLYMER

SIS-3*   20% Styrene

Polymer Added	Wt. %	Stress ($\sigma_{\lambda=4}$) Kg.Cm.$^{-2}$	T.S.(σ_b) Kg.Cm.$^{-2}$	% Set
None	0	17.5	296	25
PS(15,000)	1	21.8	291	20
PS	2	20.0	303	20
PS	5	24.0	306	30
PS	10	25.4	302	40-45
PS	20	33.5	288	50
PI(84,000)	5	14.2	263	20

* $\overline{M}_s$ = 13,700 - 109,400 - 13,700

It can be seen at once that, whereas the presence
of free polystyrene, even in large amounts, has a
negligible effect on the strength of these elastomers,
such is not the case for the presence of the diblocks.
The latter, even when present in amounts of 1-2%, have
a noticeable effect in decreasing the strength. This
indicates that any free polystyrene can quite effectively
find its way into the polystyrene block domains to form
part of the "network", and this is attested by the fact
that these films (cast from solvent) retain their
clarity (small domain size) even in presence of the
added polystyrene. Any added polyisoprene does not
have a dramatic effect, but appears to decrease the
tensile strength just as any diluent would. The diblock (SI), on the other hand, presumably leads to
"network defects", in having one free isoprene end.

In the above cases, the free polystyrene and
diblock polymers both had similar block sizes to those
in the SIS polymers. It is important to realize in this

TABLE II

EFFECT OF ADDED DIBLOCK POLYMER

SBS-5　　　21,100-63,400-21,100
SB Diblock　21,000-63,400

% Added Diblock	Stress* at $\lambda=4$	Tensile Strength*	% Set
0	46.3	319	50
1	50.3	308	50
2	47.5	264	50
5	48.3	244	50
67	14.6	49	200

* Kg.Cm.$^{-2}$

connection, that any large variation in block size may also drastically affect the entry of the added polystyrene into the domains. Thus the data in Table III show clearly the importance of the two parameters, i.e., 1) the minimum block size of polystyrene required for

TABLE III

CRITICAL VALUES FOR POLYSTYRENE BLOCK SIZE

SIS-27　7,000-60,000-7,000 (19% S)
SIS-29　5,000-80,000-5,000 (11% S)

Block Polymer	Added PS($\overline{M}_S$)		Film Props.	Stress* at $\lambda=4$	Tensile Strength*
SIS-27		-----	Clear	13	22
SIS-27	5%	6,000	Clear	18	31
SIS-27	10%	6,000	Clear	22	40
SIS-27	5%	20,000	Opaque	15	22
SIS-27	10%	20,000	Opaque	16	19
SIS-29		-----	Clear	∿ Nil	∿ Nil
SIS-29	10%	6,000	Opaque	"	"
SIS-29	10%	20,000	Opaque	"	"

93% Polyisoprene ($\overline{M}_S$ = 63,000) + 7% PS 6,000 Opaque Film
* Kg.Cm.$^{-2}$

formation of domains, and 2) the inability of a polystyrene of 20,000 M.W. to enter the domains formed by

polystyrene blocks of 7000 M.W. (The latter may be due
not to any thermodynamic incompatibility between the
polystyrene chains but to the difficulty of the longer
chain in actually fitting into the small domains.)

The "zero tensile" of SIS-29 is taken as a sign of
the absence of coherent polystyrene domains, while the
<u>increase</u> in tensile strength of SIS-27 upon addition
of the polystyrene (6000 M.W.) indicates the entry of
the latter into the domains. On the other hand, the
<u>decrease</u> in tensile strength of this block polymer upon
addition of the 20,000 M.W. polystyrene, together with
the opacity of the cast film, can be taken as proof that
this polystyrene did not enter the domains but formed a
separate phase.

The importance of polystyrene block size in formation of domains and its influence on the ability of the
blocks to actually enter such domains is also well
demonstrated in Figure 3 and Table IV. In this case,
the added polystyrene (S-30) was prepared in such a way

TABLE IV

EFFECT OF MWD OF ADDED POLYSTYRENE

Block Polymer	Wt. % Added PS	M.W. of Added PS	MWD of Added PS	T.S. (Kg.Cm.$^{-2}$)
SBS	0	----	----	127
SBS	14.9	12,500 (S-40)	Narrow	342
SBS	14.9	15,000 (S-30)	Broad	214
SIS	0	----	----	285
SIS	14.9	12,500 (S-40)	Narrow	313
SIS	14.9	15,000 (S-30)	Broad	210

M_S of SBS or SIS = 13,700 - 109,000 - 13,700

as to deliberately introduce a broader molecular weight
distribution, as shown in the Gel Permeation Chromato-
gram (Figure 3). It can be seen at once from Table IV
that this leads to a substantially lower tensile
strength in the blended polymer, presumably due to the
inability of some of the longer polystyrene chains to

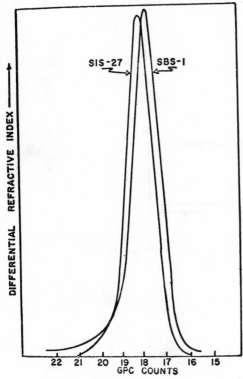

enter the domains of the block polymer. Obviously the same phenomenon could occur if the polystyrene blocks themselves had this type of broader molecular weight distribution, as a result of polymerization variables.

Acknowledgment

The experimental work described herein was carried out by Drs. J. E. McGrath, P. C. Juliano and F. C. Schwab, and Mr. C. R. Strauss. It was mainly supported by the Materials Laboratory, U. S. Air Force, under Contract AF33(615)-5362, as well as The Goodyear Tire & Rubber Co.

Figure 1. Gel Permeation Chromatographs of SBS and SIS Block Polymers, by three-stage alkyllithium polymerization.

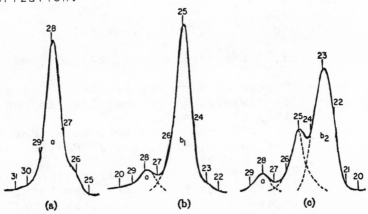

Figure 2. Gel Permeation Chromatographs[9] of (a) Poly-4-vinylbiphenyl (A block), (b) Poly-4-vinyl-biphenyl-polyisoprene (AB block),(c) Coupled AB block polymer.

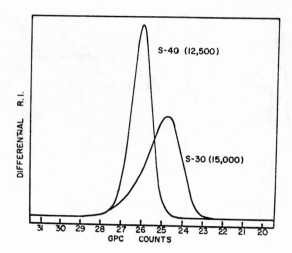

Figure 3.   Gel Permeation
Chromatograms of Polystyrene
S-30 and S-40.

Literature Cited

1.   "Block Copolymers", J. Polymer Sci., Part C, 26,
     1-209 (1969).
2.   T. L. Smith and R. A. Dickie, Ibid., p. 163.
3.   M. Morton, J. E. McGrath and P. C. Juliano, Ibid.,
     p. 99.
4.   L. J. Fetters and M. Morton, Macromolecules, 2,
     453 (1969).
5.   M. Szwarc, M. Levy and R. Milkovich, J. Am. Chem.
     Soc., 78, 2656 (1956).
6.   M. Morton and F. Ells, J. Polymer Sci., 61, 25
     (1962).
7.   R. Milkovich, South African Pat. 642,271, May 14,
     1964.
8.   M. Morton, Proceedings of the Princeton University
     Conference on Advances in Polymer Science and
     Materials, 1968 (in press).
9.   J. G. Heller, J. F. Schimscheimer, R. A. Pasternak,
     C. B. Kingsley and J. Moacanin, J. Polymer Sci.,
     Part A-I, 7, 73 (1969).
10. M. Morton, Chap. VII in "Copolymerization", G. E.
     Ham, Ed., Interscience Publishers, New York, 1964.
11. R. E. Cunningham and M. R. Trieber, J. Appl.
     Polymer Sci., 12, 23 (1968).

BLOCK COPOLYMERS OF STYRENE AND ISOPRENE:

EXPERIMENTAL DESIGN AND PRODUCT ANALYSIS

J. Prud'homme and S. Bywater

Chemistry Division, National Research Council

Ottawa, Ontario, Canada.

The use of anionic polymerization to produce block co-polymers was described some years ago[1]. In principle the copolymers can be made in a stepwise fashion, by adding each monomer in turn to the polymerization system. To produce block co-polymers as free as possible of partial polymerization products and with each block quite monodisperse requires careful planning and rigorous purification techniques. Kinetic studies made on anionic polymerization and co-polymerization are a useful guide in experimental design.

The principle requirements for good products can be summarized as follows:

(1) Careful control of polymerization conditions to exclude terminating agents.

(2) A solvent system which is not reactive and in which the carbanions are stable towards isomerization and reaction with the solvent.

(3) Rapid initiation and cross-over reactions and relatively slow propagation steps.

It should be emphasized that it will always be more difficult to produce a monodisperse block copolymer than a homopolymer of the same molecular weight. It is more difficult to remove reactive impurities from monomers than from the other reaction components. Most of the impurities can be assumed to be fast-reacting and in

11

homopolymer formation these then simply destroy initiator
or very short chains with little effect on the distribution of precipitable polymer. In formation of block
copolymers by sequential monomer addition, these impurities (except in the first step) destroy the activity
of polymer chains of appreciable molecular weight
resulting in contamination by homopolymer and other
materials of smaller block number than desired.

     In the styrene-diene systems, we are restricted to
hydrocarbon solvents and lithium alkyl initiators to
retain the desirable 1,4 microstructure of the diene.
These systems have other advantages; the initiators are
soluble, carbanion isomerizations are very slow and in
general hydrocarbons do not react with the active centres
at normal temperatures. Chain propagation is relatively
slow so that it will not be difficult to keep the reaction
mixture homogeneous in composition. Unfortunately under
many conditions, chain initiation by lithium alkyls is
often slow also. This is true for all alkyls in aliphatic solvents, but in aromatic solvents sec-butyl lithium
initiation of both isoprene and styrene polymerization
is much more rapid than for the more common n-butyl
lithium[2]. The ratio of the initiation rates for
styrene in benzene is over two hundred[3] so with sec-
butyl lithium, initiation is complete within the mixing
time at monomer concentrations commonly used for polymer
preparation. Benzene should be used rather than toluene
for the latter compound reacts with both initiator and
polystyryl lithium at a slow but noticeable rate. n-butyl
lithium could be used in presence of a small amount of
THF to accelerate the initiation rate[4] but unfortunately
in the amounts required would produce a noticeable change
in diene microstructure. The presence of less basic
ethers may be possible to speed up initiation without
too large a change in diene microstructure[5] but on the
whole, the sec-butyl lithium/benzene system offers the
best advantage both in simplicity and because of the
fact that sec-butyl lithium is easily purified by short-
path vacuum distillation. Its thermal stability is,
however, lower than for n-butyl lithium.

     The propagation reactions in benzene are relatively
slow and of the two cross-over reactions, one is fast
and the other slow. Kinetic studies[6] indicate that
the reaction of poly-styryl lithium with isoprene is
about twenty five times faster than the subsequent
propagation rate of isoprene. (These results refer to
cyclohexane as solvent but the results should be similar
in benzene.) A rapid interchange from polystyrylithium

to polyisoprenyllithium will occur when the isoprene is
added which will then polymerize slowly. A two block
styrene-isoprene copolymer of low polydispersity is thus
easy to prepare if styrene is the monomer added first,
limited only by the precautions taken to exclude impurities.
For a three block styrene-isoprene-styrene copolymer a
difficulty arises in that the reaction of polyisoprenyl
lithium with styrene is about twenty times slower than
the subsequent homopropagation rate of styrene. Hetero-
geniety in the final styrene block will be produced
unless corrective measures are taken. This can consist
of the addition of a small amount of THF with the final
styrene addition. (Earlier addition would of course
affect the microstructure of the isoprene block). This
accelerates the cross over reaction but has little effect
on the homopolymerization rate of styrene at ∿ 0.15 molar
THF (fig. 1). At this concentration the two rates are
about equal which should be sufficient to ensure that
heterogeneity of the final block will be within acceptable
limits.

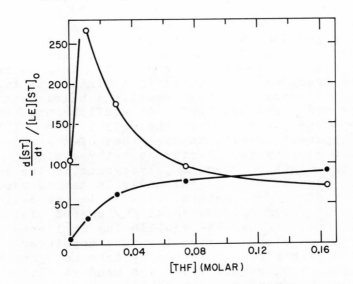

Fig. 1. Effect of THF on the propagation rate of styrene
        O and the reaction of polyisoprenyl lithium with
        styrene, ● (in benzene at 30°)

Even with careful experimental design it is desirable
to check the products, particularly if samples of high
molecular weight are required. The latter require the
use of lower catalyst levels and as these are decreased,
impurities will become important at a point depending
on the care taken in purification of monomers and solvent.
Gel permeation chromatography is perhaps the most commonly
used analytical tool. Its application has been often
described and so advantage will be taken to discuss other
methods. For instance for a styrene-isoprene-styrene
copolymer, use can be made of the ultra-violet spectrum
of polystyryl lithium. The optical density at $\sim$ 334 mμ
at the end of the reaction should correspond to that
observed during the polymerization of the initial
styrene block. If it is lower, some termination has
occurred. (Correction is required for volume changes
and for the change in extinction coefficient if THF
has been used at the second cross-over stage). For a
styrene-isoprene two block copolymer this technique
is not so useful since the polyisoprenyl lithium spectrum
is half-obscured by benzene absorption. A better
knowledge of the extinction coefficient of polyisoprenyl
lithium at longer wavelengths would make this method
valuable.

Density gradient ultracentrifugation[7] appears to
be a powerful method of product analysis, particularly
for copolymers of high molecular weight. In this method,
the density gradient is produced in a mixture of two
solvents of different density (usually a halogenated
hydrocarbon and a hydrocarbon). At equilibrium, each
species separates as a band in the cell at the point
where its apparent density matches that of the solvent.
The order of density is polystyrene > three block
polymer > two block polymer, i.e. polystyrene is the denser
component. Because the optical analysis method used
gives a plot of dc/dx against position in the cell
(x), a band of polymer produces an $\bigwedge$ shaped plot on
the plate. Fig. 2 shows the equilibrium Schlieren
pattern of a two block styrene-isoprene copolymer
(50/50) in a carbon tetrachloride/cyclohexane gradient.
A spinco model E ultracentrifuge was used at 50,740
r.p.m. A 50/50 v/v mixture of solvents was chosen so
that the block copolymer floats and the polystyrene
impurity comes to equilibrium at about the centre of
the cell. The integrated area corresponds to about 9%
polystyrene from which it will be seen that 1% polystyrene can be determined under these conditions. At a
higher total polymer concentration the method could be

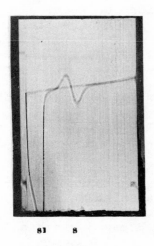

Fig. 2. Schlieren diagram of
an impure two block poly-
mer, represents
polystyrene. Two block
polymer at left near
meniscus.

field ⟶

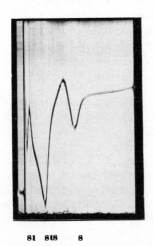

Fig. 3. (left) Schlieren diagram of an impure 3-block
copolymer. Polystyrene near cell centre, 2 and 3
blocks at left near meniscus.

Fig. 4. (right) Same polymer in a different gradient.
2 block impurity at left, other components towards
cell bottom.

even more sensitive. Figs. 3 and 4 show the equilibrium
patterns of a 3 block styrene-isoprene-styrene co-polymer.
By adjusting the solvent composition the required
impurity can be brought to the centre of the field. In
figure 3 the solvent composition was 50 (v/v) % CCl_4-
cyclohexane; the two and 3 block copolymers float leaving
a band of polystyrene at the cell centre. In figure 4
the solvent composition was 36.4% CCl_4 and the two block
copolymer appears at near cell centre, the rest sinks.

The ability to observe separate bands for each
component depends on diffusion being slow and thus the
method is most valuable for products of quite high
molecular weight and is probably best regarded as
complementary to gel-permeation chromotography where
the resolution tends to decrease at high molecular weights.
The products illustrated had molecular weights ∿500,000
and ∿900,000 respectively for the 2 and 3 block polymers.

The ultracentrifuge can also be used in the normal
sedimentation velocity mode for product analysis. The
separation achieved on the above samples was, however
not as good as with the density gradient technique. A
number of solvents was investigated with disappointing
results. Only in methyl isobutyl ketone (close to a
Θ solvent at room temperature for both polyisoprene and
polystyrene) was a reasonable separation achieved. This
method, although much more rapid than is the density
gradient technique, has the added disadvantage that
sedimentation velocity is a function of concentration,
molecular weight and composition. The identification
of individual peaks is therefore more difficult and
their order may concievably change even with sample
molecular weight. In contrast in a given density
gradient each component can be easily identified, as
it appears at a fixed position independent of the molecular weight of the sample.

References

1. M. Szwarc, M. Levy, and R. Milkovich,   J. Am. Chem.
       Soc., 78, 2656 (1956).

2. H.L. Hsieh,   J. Polymer Sci., A3, 163 (1965).

3. S. Bywater and D.J. Worsfold,   J. Orgmet. Chem., 10,
       1 (1967).

4. S. Bywater and D.J. Worsfold, Can. J. Chem., 40 1564
    (1962).

5. A.V. Tobolsky and C.E. Rogers, J. Polymer Sci., 40 ,
    73 (1959).

6. D.J. Worsfold, J. Polymer Sci., A-1:5, 2783 (1967).

7. M. Meselson, F.W. Stahl and J. Vinograd, Proc. Nat.
    Acad. Sci., 43,581 (1957).

SMALL ANGLE X-RAY SCATTERING STUDIES

OF TRI-BLOCK POLYMERS

D. McIntyre[a] and E.Campos-Lopez[b]

(a) Institute of Polymer Science

The University of Akron, Akron, Ohio 44304 and

(b) Facultad de Quimica Universidad Nacional

Autonoma de Mexico

INTRODUCTION

Tri-block polymers of styrene-butadiene-styrene(SBS) and styrene-isoprene-styrene(SIS) phase separate into domains of quite different average composition and rather large size. The microphases of good elastomeric materials consist of a single phase (styrene rich) of spheres imbedded in a matrix of a second phase (diene rich). The average dimension of these spherical phases is greater than 100Å. The electron microscopic work of Beecher, Marker, Bradford and Aggarwal [1] has shown the large effect that the casting solvent can have on the morphology of the tri-block cast film. More recent electron microscopic work of Matsuo [2] has shown how a change of casting solvent can change the domain structure in the plane perpendicular to the film relative to the domain structure seen in the plane parallel to the flat film.

Domain sizes of the order of 100Å are ideally suited for small angle x-ray scattering studies. Furthermore, the small angle x-ray scattering measurements complement the electron microscopic studies in the following ways: (1) All domains throughout the sample are averaged so that surface effects or unrepresentative sampling are avoided. As a consequence, an average size of the domains can be easily determined, provided that the Guinier approximation can be applied. (2) The

19

average spacing of the domains from each other can be determined, as well as the statistical variation of the spacing. (3) Under ideal circumstances the composition of the domains can be estimated. (4) The change of the morphology with variation of stress, temperature, or pressure is easily studied.

This study briefly describes the study of the influence of three variables; casting solvent, added homopolymer, and change of the molecular weight of the end block-polymer on the interparticle scattering peak. This scattering peak is easily accessible to small angle x-ray scattering measurements and enables the perfection of two-phase structure to be determined. Finally, this study reports measurements of the scattering from within the domains rather than between the domains. From measurements at relatively high angles ($2\theta > 1000$ sec) a domain size may be determined. Below the main inter-particle peak ($2\theta < 1000$ sec) even more information about the spatial arrangement of the domains may be determined. The latter studies will only be briefly discussed.

The study of any material by a scattering technique is made much more revealing, if a simple model can be assumed in the interpretation of the scattering measure-ments. As more and more scattering measurements are made with substitutional atoms, fixed orientations, and ab-solute intensities, it is possible to gain a detailed knowledge of the material. In this study it is assumed that the electron microscopic picture of spherical domains for solvent-cast films is correct, when the domain size is determined. Below are the main equations describing the scattering from a two-phase system of differing electron density, when one phase is spheres.

Guinier [3] has shown that the scattering of randomly oriented domains in concentrated solution decreases the scattering in the direction of the incident beam and finally increases the scattering at another larger angle until a maximum appears at high concentra-tions. The position of this peak can be used to calcu-late a characteristic distance by the Bragg relation, $\lambda = 2d \sin \theta$, where 2θ is the scattering angle.

The total scattering, represented as $I(s)$ in eqn (1), is composed of the single particle scattering, $F^2(s)$, and the single particle interference. The latter interference is related to the geometrical positions of the particle centers and is given by the correlation function, $P(r)$; s is the scattering vector of magnitude $2 \sin \theta/\lambda$, x is the distance, and v is the volume of the particle.

$$I(s) = F^2(s) \left[1 + \frac{2}{sv_1}\right] \int_0^\infty [P(x)-1]x \sin(2\pi sx) \, dx \quad (1)$$

$$F^2(s) \doteq \left[(\rho-\rho_0) \frac{4}{3} \pi a^3\right]^2 \frac{9(\sin 2\pi as - 2\pi as \cos 2\pi as)^2}{(2\pi as)^6} \quad (2)$$

Eqn (2) describes the scattering of uniform spheres of radius a and electron density ρ immersed in a medium of electron density ρ_0. The scattering intensity calculated from eqn (2) is a series of maxima and minima when plotted against s.

In the study of tri-block polymers of elastomeric interest the interparticle interference is always present. Also the maximum value of electron density differences is fixed by the choice of homopolymers, although it may not be attained in a given sample. Therefore the intensities of domain scattering may be very weak and necessitate long counting times.

EXPERIMENTAL

Polymers

Preparation - Tri-block polymers (S-B-S, S-I-S) of butadiene(B) and styrene(S), and isoprene(I) and styrene(S) were made anionically by high vacuum techniques as generally described by Morton [4], Schwab [5] and Fetters [6].

Characterization - The overall molecular weights of tri-block and homopolymer were determined by osmotic pressure measurements; the percent styrene was calculated from the reaction stoichiometry. The chemical and physical characterization is described in references [4 and 5].

Solvents - ACS Reagent grade carbon tetrachloride, benzene, heptane, tetrahydrofurane (THF), and methyl ethyl ketone (MEK) were used in the film casting.

Film Casting

The polymer casting solutions were made by dissolving enough polymer by shaking for a day to make a 10% solution. The films were cast at room temperature on glass with a Gardner film casting knife set for a 2 mil thickness. The films were then dried under high vacuum at 95^0C until they came to constant weight. The casting solvents were THF-MEK (90%THF-10%MEK), benzene-heptane (90%benzene-10%heptane), and CCl_4.

X-Ray Measurements

A Hart-Bonse[7,8] low angle x-ray camera manufactured by Advanced Metals Research, Burlington, Mass., was used. The sample thickness was fixed at 1.83 mm and consisted of several layers of cast film. The transmission coefficient was approximately 0.5. The generator was the ultra-stable Philips Model PW 1310. The exposure time for each scattering angle varied. For scattering angles four 0-1000 sec readings were taken every 20 sec of arc. Counting times were 100 sec. At scattering angles greater than 1000 sec the intensity was so low ($\sim 3c/_s$) that counting times of 25 minutes were required for each point. (All data are plotted with intensity as counts per 100 sec, or 1000 sec.)

The corrections for background were made by measuring the transmission coefficient of the sample directly at $0°$ with attenuation filters in the beam.[9] The camera can routinely allow scattering to be measured within 20 sec of the incident beam.

RESULTS

Solvent Effects

A sample of tri-block polymer, polystyrene-polybutadiene-polystyrene of approximately 40% styrene and of molecular weight 21,100 - 63,400 - 21,100 was cast as a film from different solvents, then dried.

The scattering patterns up to 1200 sec of arc are shown in Figure (1). The scattering pattern up to 300 sec is characteristic of the single domain scattering that has been distorted by interdomain interference. The Bragg spacings interparticle interference are shown in Table I.

Table I

Bragg distances from Different Casting Solvents

(S-B-S (21,100-63,400-21,100) - 40% styrene)

CCl_4 530Å

THF/MEK   430Å

Benzene/Heptane   410Å

Variation of End Block Polymers, Middle Block Fixed

S-B-S tri-block polymers having a fixed B block of molecular weight 63,400 and variable S blocks were studied as films cast from THF/MEK mixtures. The x-ray scattering results are shown in Figure (2). Not only is

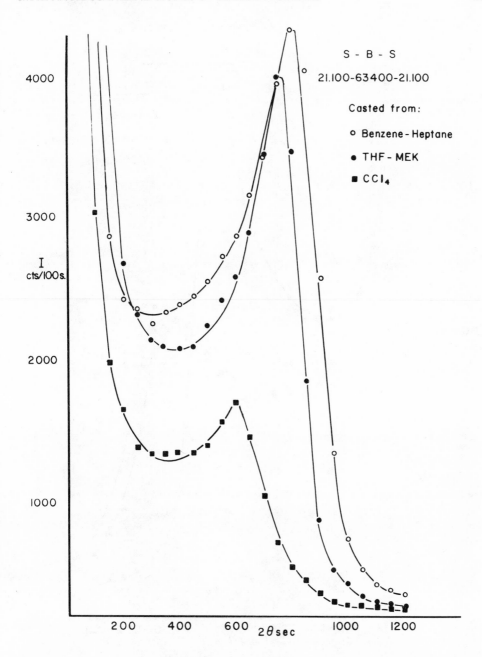

S - B - S

21.100-63400-21.100

Casted from:

○ Benzene-Heptane

● THF - MEK

■ CCl$_4$

Figure 1   Scattering from S-B-S Polymer Cast from
Different Solvents

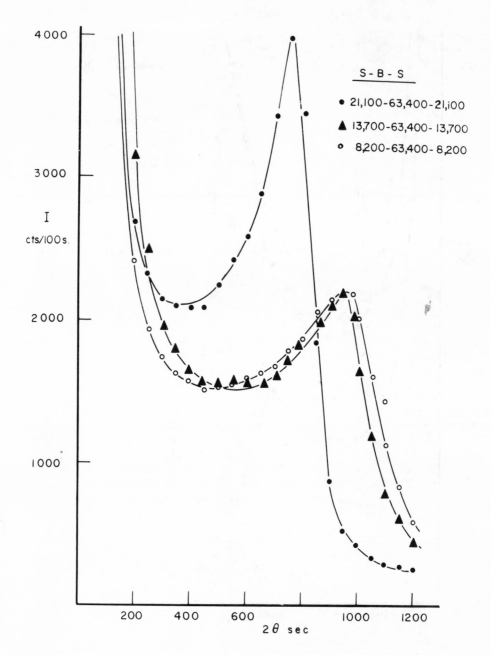

Figure 2   Scattering from S-B-S Polymers Having
                    Different Composition

the S block molecular weight changing, but also the volume per cent of styrene is decreasing as the end block molecular weight is decreased. Table II lists the Bragg spacings for the data of Figure (2).

TABLE II

Bragg Distances from Different S-B-S Polymers

MW (S-B-S)	% Styrene	Bragg Distance($\overset{o}{A}$)
21,100-63,400-21,100	40	430
13,700-63,400-13,700	30	339
8,200-63,400- 8,200	20	325

Homopolymer-Tri-Block Polymer Blends

Homopolymer of polystyrene (MW 15,000) was added to the THF casting solution of the tri-block polymer (S-I-S 14,000-109,000-14,000) in varying amounts [5]. The films were examined by x-ray scattering and the data are plotted in Figure (3). The Bragg distances vary by 30%, being larger the greater the amount of homopolymer added. The half-width of the peak also narrows to a minimum at 5% and then broadens at greater compositions of blended homopolymer.

Intraparticle or Intradomain Scattering

Figure (4) shows the scattering of an S-B-S polymer of molecular weights 21,100-63,400-21,100, respectively. Only the scattering beyond 1000 sec of angle is shown. The film was cast from THF/MEK and may be compared, at very small angles, with the curve shown in Figure (2). The counts are now reduced by a factor of ten. The vertical lines are indicative of the experimental error at each angle. Table III lists the data from Figure (3) that must be substituted into eqn (2) to determine the radius of homogeneous, isolated spheres. The various maxima that are assigned to the spheres must yield the same radius if the assumptions of eqn (2) are true and the orders of the maxima are properly assigned.

TABLE III

Sizes of Domains as Radius of Spheres from Eqn (2)

Maxima	2θ	sa	a ($\overset{o}{A}$)
0	0	0	−
1	1500	5.31	179
2	2450	8.63	178
3	3550	11.85	170

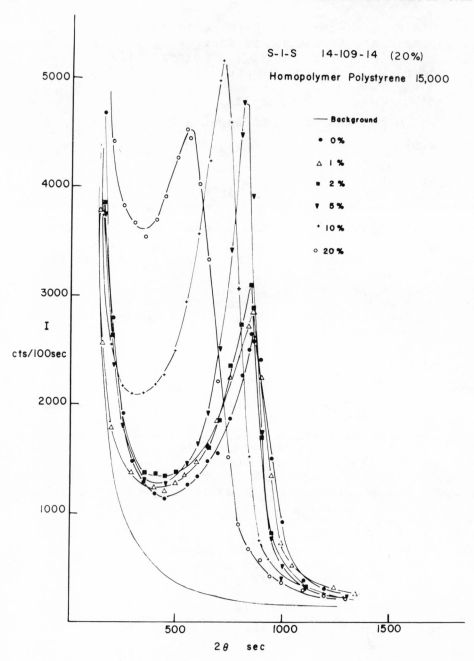

Figure 3   Scattering from S-I-S Polymer with Blends of Homopolymer

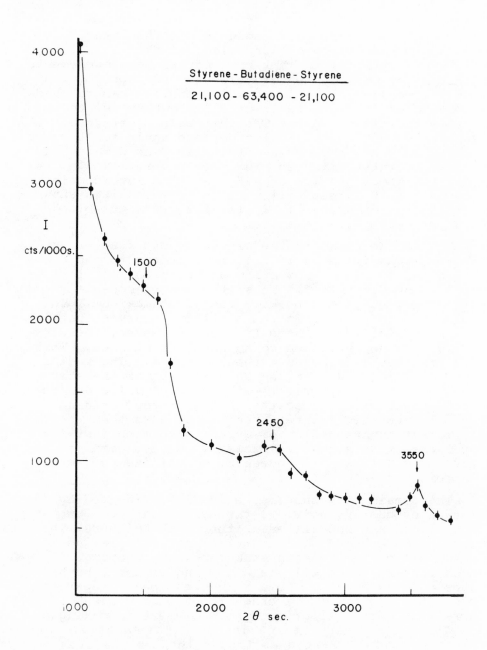

Figure 4   Scattering of S-B-S Polymer at Higher
                  Angles

DISCUSSION

The x-ray experiments conducted in this study demonstrate the wealth of morphological detail that may be obtained by small angle x-ray scattering measurements. Earlier x-ray work of Hendus, Illers, and Ropte [10] on block polymers, using a low resolution film camera, had only indicated the existence of a broad band due to interdomain interference. The scattering pattern of S-B-S (21,000-63,400-21,100) from zero angle to 4000 sec shows clearly how the radius of the styrene domains and the average interdomain spacing may be determined. The calculated domain radius of ~180Å and the interdomain distance of 430 agrees well with the domain sizes shown by electron microscopy [1, 2, 11] for other tri-block S-B-S polymers.

To fit the peaks in the single sphere scattering region by an assignment of different orders to the peaks at 2450 and 3550 will now allow a consistant scattering radius to be determined. Thus it seems reasonable to assign the 1500 second peak to the first maximum. There is much fine structure in the scattering curves that is not easily visible in these figures. However, these details have not yet been so uniquely identified. Work is in progress to elucidate these details. The radius of gyration deduced from the Guinier plot of the scattering data is in agreement with the 180Å value deduced by the single particle scattering. But the influence of the interparticle interference increases the uncertainty in the radius of gyration so much that its usefulness at this time is minimal. A more complete theory of scattering is necessary to utilize this easy measurement.

The domain structure is decidedly influenced by the type of casting solvent, the volume per cent of the first block, and impurities in the block polymers. The change of peak height in Figure (1) for CCl_4 is due to a less well-defined domain structure. This change is not surprising since CCl_4 is a good solvent for both phases and may not lead to a very clean phase separation. The shift of 20% in the average interdomain distance may represent the use of both homopolymers in making a more diffuse boundary layer. The diffuseness allows the domains to scatter x-rays as though the domains are larger in extent. An analysis of peak width could lead to some information on the irregularity of the domain spacing, just as the breadth of the intradomain peak can lead to statements of the uniformity of the sphere sizes. However, it is not clear yet how to analyze the curves in an accurate manner. If it is simply necessary to take the width of the interdomain peak at a height a given fraction of the distance above the minimum, the THF/MEK system would be judged the most ordered. The benzene-heptane system would then be considered to yield the most completely separated phases since the peak height is higher. Whether a 7 to 10% increase of absolute intensity at uniform film thickness is meaningful is not presently known.

The change of interdomain distance on increasing the styrene content of S-B-S from 20% to 40% at fixed butadiene molecular weight is also striking. A 30% increase of average interdomain spacing and an increase of the peak height by a factor of two suggests several possible explanations. (1) There is little morphological change on increasing the styrene content from 20% to 30% – even though the actual domain scatterer (styrene) has increased by 50% over the original 20%. Yet on increasing the styrene content to 40% the peak intensity almost doubles. These results might indicate that contrary to earlier theoretical work of Meier [12] there are changes in the completeness of phase separation at molecular weights up to 15,000 of polystyrene. That there are phase separations is unambiguous due to the interdomain peak. At even lower molecular weights of end block the phase separation must be incomplete since the interdomain scattering peak disappears. Mechanical measurements of Morton [4] suggest a loss of domain structure.

The homopolymer blending experiments had to be carried out on a 20% styrene tri-block polymer of S-I-S. Unfortunately a comparison can not be made with the other S-B-S x-ray data. Up to a 5% addition of homopolymer, and very dramatically at 5%, the intensity increases until it is almost double. The angular position of the peak shifts no more than 7%. It appears that the domains have become much more separated in composition upon the homopolymer addition, probably due to the viscous and thermodynamic affect of the homopolymer at the domain boundaries. Further additions of added homopolymer does nothing but shift the peak distance ever closer to zero angle and broaden the curve. Eventually the homopolymer at only a 20% addition (or a total of 40% styrene) scatters like no other simple tri-block system. It appears as though the styrene interdomain distance might be within 10% of an ordinary 40% styrene tri-block, but the regularity of the domain spacings has begun to be lost. Perhaps even separate very-enriched styrene homopolymer microphases are present.

    More work is presently being carried out on block polymers throughout the entire angular range of low angle x-ray scattering to make the interpretation of the low x-ray scattering easier and more rapid. This method is perhaps the necessary complement to the electron microscope that is required to elucidate the tri-block morphology completely.

Acknowledgement

    The authors thank Drs. M. Morton and F. Schwab, and C. Strauss for making their polymer samples available to us and thank N. Rounds for his help in this work. E. Campos-Lopez wishes to thank the Instituto Nacional de la Investigacion Cientifica for their support during his stay at the Institute of Polymer Science in 1969. D. McIntyre wishes to thank the Office of Saline Water for partial support of this work.

REFERENCES

1. J. F. Beecher, L. Marker, R. D. Bradford, and S. L. Aggarwal, J. Poly. Sci., $\underline{C}$, $\underline{26}$, 117 (1969)

2. M. Matsuo, Japanese Plastic, $\underline{2}$, ^{3}, 6 (1968)

3. A. Guinier, Small-Angle Scattering of X-rays, Wiley, (1955)

4. M. Morton, This Symposium

5. F. Schwab, Ph.D. Thesis, Univ. of Akron, 1969

6. L. Fetters, J. Polymer Sci. $\underline{C}$, $\underline{26}$, 1 (1969)

7. U. Bonse and M. Hart, Appl. Phys. Letters, $\underline{7}$, 238 (1965)

8. U. Bonse and M. Hart, Z. Physik, $\underline{189}$, 151 (1966)

9. M. E. Myers, M.S. Thesis, Univ. of Akron, 1969

10. H. Hendus, K. H. Illers, E. Ropte, Kolloid Z, Polymere, $\underline{216/217}$, 110 (1967)

11. P. R. Lewis and C. Price, Nature, $\underline{223}$, 494 (1969)

12. D. J. Meier, J. Polymer Sci., $\underline{C}$, $\underline{26}$, 81 (1969)

RELATIONSHIP OF MORPHOLOGY TO PHYSICAL PROPERTIES OF STYRENE-BUTADIENE BLOCK COPOLYMERS

G.A. HARPELL AND C.E. WILKES

B.F. GOODRICH RESEARCH

INTRODUCTION

Block copolymers of the type polystyrene-polybutadiene-polystyrene (SBS) behave like reinforced vulcanized elastomers at room temperature.[1-3] The addition of low molecular weight polystyrene adds further reinforcement. (For example, this addition to a SBS polymer of 30 wt. % styrene to give an overall composition of 40 wt. % styrene increased the modulus at an extension ratio of 5 from 23 to 44 kg./cm.^{2} at a strain rate of 31.5 min.$^{-1}$.)

This is an ideal model system for studying the relationships between the morphology and physical properties of reinforced elastomer systems. It is particularly valuable since the reinforcing domain sizes can easily be controlled[4] and the domain shapes can be varied from spherical to rod-like to lamellar[5,6] for similar block copolymers.

The polystyrene end-blocks segregate to form the reinforcing, crosslinking domains. Since the polystyrene and polybutadiene block are chemically connected, segregation into pure domains is impeded. Flow at elevated temperatures, such as occurs in milling or extrusion operations, and annealing will both influence the sample morphology. We have determined how

these changes in domain structure with thermal history affect
the torsional and tensile moduli and extensibility. Since these
block polymers show very discreet low-angle x-ray diffraction
patterns[6] we used this technique to follow changes in morphology
with annealing conditions.

This report is limited to a study of a blend of tri-block SBS
copolymer containing 30 wt. % styrene with enough polystyrene
homopolymer to give a total of 40 wt. % styrene.

EXPERIMENTAL

The polystyrene-polybutadiene-polystyrene block copolymer
was prepared in a 5 gallon stirred reactor previously purged with
a living polystyrene solution. Impurities in the styrene and
benzene solvent were titrated with secondary butyllithium (using
a specially installed sight-glass and pump). On the first appearance of color an appropriate additional amount of secondary
butyllithium was added. After the polymerization of the styrene,
butadiene (previously distilled from secondary butyllithium) was
transferred into the reactor through lines previously flushed with
living polystyrene solution. After the butadiene polymerization
was complete, a coupling agent was added to the living two-block
polymer to form the three-block polystyrene-polybutadiene-
polystyrene tri-block polymer. The coupling reaction was approximately 80% quantitative as determined by gel fractionation
chromatography. The polystyrene homo-polymer was prepared
in a similar manner without coupling.

The 30 wt. % styrene block copolymer (M_n of polystyrene
end blocks = 16,500) and polystyrene homopolymer (M_n = 7,600)
were dissolved in benzene and films were cast by slow evaporation. The films were further dried in a vacuum oven at 50°C for
16 hours. They were then milled on a 4" rubber mill at approximately 140°C. Sheets were cut from the milled stock and molded
in a constant volume mold.

Stress-strain measurements were obtained on ring samples
with 1" inside diameter and cross-sectional area of 0.050" by
0.050". Torsional moduli were measured by the Gehman Low
Temperature Torsional Modulus Test (ASTM Designation D1053-
65). Low angle X-ray diffraction patterns were obtained using
a Rigaku Denki RU2 rotating anode generator and goniometer and

slit collimation. Peaks up to ~600A° were quite resolvable. No slit height corrections were made on the observed spacings.

RESULTS AND DISCUSSION

Low-Angle X-Ray Diffraction

Fig. 1 illustrates that the well annealed sample (Sample No. 5, molded for 82 minutes at 168°C) gives the most intense and narrow interdomain scattering peak at the highest scattering angle. (The interdomain distance decreases as the scattering angle 2 Θ increases). This indicates a high degree of regularity in the interdomain distance. Less effective annealing (Sample No. 1, molded for 18 minutes at 112°C) causes the peak to be broader and at a smaller scattering angle. This indicates that the range of distribution in interdomain distance is broader and that the average interdomain distance is also larger. In the unmolded sample only a broad shoulder appears, indicating much less regularity of interdomain spacing. We believe that the increase in regularity of interdomain distance with annealing is accompanied by an increase in domain purity, i.e., occluded polybutadiene is expelled from the polystyrene domains.

The intense low-angle x-ray peaks were due to interparticle interference and not related to particle size. We found, as did Hendus and coworkers,[6] that when we stretched the films, the D-spacing along the stretch direction increased and perpendicular to stretch direction it decreased. Since, at very low elongations (5-50%) the polystyrene domains were not likely to deform markedly, the observed changes in x-ray spacings must have been due to changes in the interparticle distances. To date, we have not been able to extract particle size information from the diffraction curves.

Partial pole figures (Fig. 2a and 2b) for the well annealed sample (5) show the variation of the intensity of the interparticle scattering maximum with orientation. The variation of the scattering intensity with χ (sample surface kept normal to the x-ray beam) shows that the sample is highly oriented with maximum order perpendicular to the mill direction. Varying the angle between the sample surface and the x-ray beam ($\emptyset$) results in a modest increase in intensity at plus and minus angles to perpendicular. Most of this increase in sample

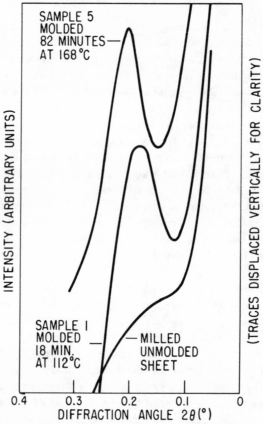

Fig. I. LOW ANGLE X-RAY DIFFRACTION
CURVES (Sample Mill Direction Parallel
to X-Ray Slits)

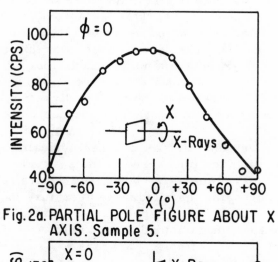

Fig. 2a. PARTIAL POLE FIGURE ABOUT X AXIS. Sample 5.

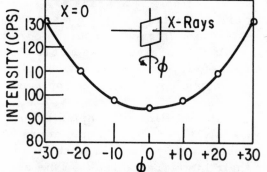

Fig. 2b. PARTIAL POLE FIGURE ABOUT ϕ AXIS. Sample 5.

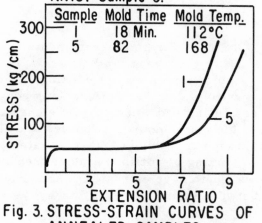

Fig. 3. STRESS-STRAIN CURVES OF ANNEALED SAMPLES

scattering can be accounted for by the increase in effective path
length. If the sample consisted of lamellae normal to the surface,
a sharp decrease in intensity would be observed as $\emptyset$ is increased
or decreased from $\emptyset = 0$. The data suggests that this system
contains rodlike polystyrene domains with their long axes parallel
to the mill direction. An alternative interpretation is that the
domains are lamellar, parallel to the milling direction, but randomly oriented with respect to the sample surface.

Table 1 includes the effect of different annealing conditions
on the interdomain distance (measured at the peak maximum--
molded surface normal to the x-ray beam--slits parallel to the
mill direction). Increasing the molding temperature or time
causes the interdomain distance to decrease. The interdomain
distance for the milled but unmolded stock is 565 A°. The interdomain distance of 415 A° (sample 5) appears to be a lower
limiting value, and more extensive annealing does not give rise
to a further decrease. A sample annealed for 8 hours at 140°C
showed a D-spacing of 415 A°. This is reasonable since polystyrene end-block length and overall composition will dictate
limits to variations in morphology with annealing conditions.

We have found that cast films of coupled three block polystyrene-polybutadiene-polystyrene block copolymers and their two
block polystyrene-polybutadiene precursors exhibit the same
scattering angle for their interference peak. This suggests that
minor amounts of two-block polymer impurity will not significantly
affect the polymer morphology.

Stress-Strain Measurements

On elongation, the samples exhibit necking, which is complete by an extension ratio of 4 or 5. This necking suggests that
the polystyrene phase is quasi-continuous. On elongation, bridge
structures connecting the major domain particles are disrupted.
The modulus at low extension ratios is that of a material in a
meta-stable state, consisting of portions almost unchanged in
dimensions and others which have elongated considerably--with
a transition region between them.

At higher elongations, where necking is complete, samples
molded for longer times at higher mold temperatures exhibit
lower moduli than samples molded for shorter times at lower

TABLE I

EFFECT OF MOLD CONDITIONS ON MORPHOLOGY AND PHYSICAL PROPERTIES

Sample Number	Mold Time (min)	Mold Temp (°C)	Interdomain Distance (A)	STRESS-STRAIN PARAMETERS*				Torsional Modulus Value $(180-\Theta)/\Theta$ at -60°C
				MOD(5) (kg/cm^2)	MOD(8) (kg/cm^2)	MOD(BREAK) (kg/cm^2)	Ex.Ratio BREAK	
1	18	112	525	37.2	188.6	274.2	8.75	1.9
2	82	112	495	36.6	175.3	284.0	9.01	2.0
3	50	140	440	35.7	149.6	275.6	9.11	2.8
4	18	168	440	37.2	133.7	269.6	9.22	3.2
5	82	168	415	37.6	105.0	248.0	9.50	3.8

*Strain rate = 0.631 min^{-1}. MOD(N) corresponds to the modulus at an extension ratio of N.

temperatures. This is clearly shown in Fig. 3. Stress-strain data in Table 1 show that if the mold temperature is held constant (either at 112 or 168°C) both the modulus at an extension ratio of 8 and the interdomain distance decrease with molding time, while extensibility increases. Similar results are obtained if the mold time is held constant and the mold temperature is increased.

Good annealing conditions promote the expulsion of occluded polybutadiene from the polystyrene domains. This gives rise to a lower apparent crosslink density and thus a lower modulus. Disruption of the domains and rupture of the sample occurs at slightly higher elongations.

Gehman Torsional Moduli

Fig. 4 shows the variation of torsional modulus with temperature for samples prepared under different molding conditions. The deformation of the sample in this test is very small and extensive disruption of the polystyrene is unlikely.

The torsional modulus increases with increasing mold time and temperature. Data in Table 1 show that, as annealing proceeds, the torsional modulus increases and the interdomain distance decreases. Thus, extensive annealing causes the quasi-continuous polystyrene phase to become purer and therefore stronger.

CONCLUSIONS

The structure of the solution cast, milled, molded blend of tri-block SBS copolymer and polystyrene is pictured as a quasi-continuous, highly oriented, para-crystalline arrangement of rod-like or lamellar polystyrene domains in a continuous polybuta- diene matrix. During the film casting operation, phase segrega- tion occurs. Hot milling causes some disruption of the domain structure and some remixing of the phases. It is at this stage

Black areas indicate Polystyrene domains

← Mill Direction →

that the high degree of orientation is introduced. Annealing at

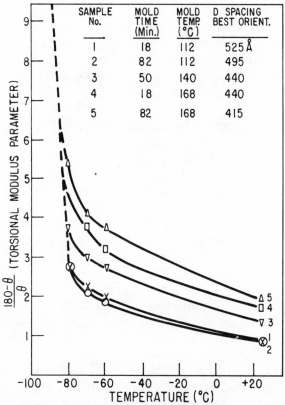

Fig. 4. TORSIONAL MODULUS OF BLEND OF SBS (30 wt. % STYRENE) PLUS POLYSTYRENE
Overall Composition — 40 wt. % Styrene

temperatures well above the glass transition temperature of
polystyrene results in increased regularity in the para-crystalline
arrangement of the domains. It also decreases the interdomain
distance. Thus, upon annealing, the domains appear to contract
in size due to the expulsion of occluded polybutadiene. This is
quite unlike the case of crystalline polymers -- where the domains
(lamellae) usually increase in size (thickness) with annealing at
elevated temperatures. It should be pointed out that, for low
angle x-ray scattering from very long rods or very extended
lamellae, information is limited to the lateral packing distance
(and in some cases the thickness). Therefore, we can say nothing
about the length or longitudinal size of these domains or their
separation along the alignment direction.

It appears paradoxical that with annealing, the polystyrene
domains become smaller and at the same time closer together.
A consequence of this is that the gross dimensions of the sample
would be expected to change. Indeed, upon heating the samples
at various temperatures (above 100°C), we observed extensive
shrinkage along the original milling direction and increase in
thickness. (We observed shrinkages up to 30% in times well under
1 minute, depending on the annealing temperatures). The polybutadiene chains connecting the oriented polystyrene domains in
the milled stock must have been in a state of considerable extension. Heating the samples above the T_g of polystyrene freed the
polybutadiene trapped in the polystyrene domains. We feel that
the attainment of the more nearly random-coil configuration by
the polybutadiene chains (relieving the stress) was accompanied
by some movement of the polystyrene domains. They must have
been pulled in closer together and better aligned at this time.
Although the annealings were done in a constant volume mold,
most of the reorganization occurred during the 30 seconds the
samples were held at the particular temperature before pressure
was applied.

This picture of the structure is consistent with the increase in
torsional modulus at low effective elongations, the decrease in
tensile modulus at high elongations and the increased extensibility
which is observed with progressive annealing. None of the
samples examined exhibited gel formation on molding. Therefore,
the presence of gel may be discounted as an explanation for
changes in physical properties with molding conditions. This
preliminary model will be refined by further critical low-angle
x-ray experiments and by electron microscopy.

ACKNOWLEDGEMENTS

We wish to take this opportunity to thank Case-Western Reserve University for the use of the low-angle x-ray equipment and Professor P. Geil and A. Burmester for helpful discussion.

REFERENCES

1. E. Fischer and J. F. Henderson, Preprint, International Symposium on Macromolecular Chemistry, Toronto, A10.5 (1968).

2. N. V. Mautschappij, Shell International Research, Brittish Patents 1, 000, 090 (1965) and 1, 035, 873 (1966).

3. M. Morton, J. E. McGrath, and P. C. Juliano, Symposium on Structure and Properties of Elastomers, Division of Rubber Chemistry, American Chemical Society, Montreal, Canada, May (1967).

4. E. B. Bradford and E. Vanzo, J. Polymer Sci., A-1, $\underline{6}$, 1661 (1968).

5. T. Inoue, T. Soen, T. Hashimoto and H. Kawai, Preprint, International Symposium on Macromolecular Chemistry, Toronto, A10. 8 (1968).

6. H. Hendus, K. H. Illers and E. Ropte, Kolloid-Z. Polymere, 216-217, 110 (1967).

7. Structure-Property Relations of Styrene-Butadiene-Styrene Block Polymers", P. C. Juliano, PhD. Thesis, University of Akron June (1968).

MULTIPLE GLASS TRANSITIONS OF BLOCK POLYMERS. II. DIFFERENTIAL
SCANNING CALORIMETRY OF STYRENE-DIENE BLOCK COPOLYMERS

R. M. Ikeda, M. L. Wallach* and R. J. Angelo

E. I. du Pont de Nemours and Company, Inc.

Film Research Laboratory, Wilmington, Delaware

INTRODUCTION

Earlier we reported on the synthesis, dilute solution characterization and torsion pendulum analyses of a series of ABA type
styrene-diene block polymers which exhibited narrow molecular
weight distributions and sharp block segments[1,2]. The two phase
nature of the styrene-diene block copolymers was evident from two
distinct glass transitions (Tg) associated with the individual
segment types. We now report the application of differential
scanning calorimetry (DSC) to investigate the multiple Tg behavior
of these block copolymers.

EXPERIMENTAL

Polymer Preparation

The polymers were synthesized in tetrahydrofuran by Szwarc
anionic "living polymerization" techniques employing sodium
α-methyl styryl initiation. We have previously reported[2] the
experimental details.

Polymer Characterization

The series of polymer samples employed in the current calorimetry investigation are the same polymers which we previously
characterized[2].

*Present address, E. I. du Pont de Nemours & Co., Inc., Photo
 Products Department, Experimental Station Laboratory,
 Wilmington, Delaware

Infrared and nuclear magnetic resonance analyses have been
reported. Dilute solution measurements involving viscometry,
osmometry, ultracentrifugation and light scattering techniques
were also described.

Sample identification here is the same as reported previously.
The block copolymers are denoted S/B, S/I, S/B/I, etc., where the
letters identify the block segments; the first letter indicates
the center segment of the ABA type block polymer (e.g., S/B/I
represents a block terpolymer with polymer segments, Isoprene-
Butadiene-Styrene-Butadiene-Isoprene).

Table I lists the polymer samples and pertinent characterization data.

Table I

Polymer Characterization Data

Styrene-Butadiene

Polymer Designation	Mole % S*	$[\eta]_{benzene}$ dl/g	$M_n \times 10^{-5}$	M_w/M_n
S	100	0.49	0.609	1.17
S/B$_1$	72	0.71	1.05	-
S/B$_2$	55	0.76	1.02	1.22
S/B$_3$	43	0.82	0.845	-
S/B$_4$	36	1.06	1.28	-
S/B$_5$	19	2.85	3.14	-
B	0	1.19	0.914	1.08
B/S	78	1.69	1.85	-

Isoprene-Styrene-Butadiene

I	0	1.21	-	-
S/I	51	1.02	1.11	-
S/B/I	53	1.60	1.43	-
B/S/I	50	1.08	0.778	-

*NMR analyses

Differential Scanning Calorimetry (DSC) Analyses

The Du Pont 900 Differential Thermal Analyzer with the DSC
cell was employed.

Polymer specimens (20 to 30 mg.) were compressed into discs
at 25 °C. before inserting into the DSC cell. An equivalent amount
of powdered Vespel® polyimide was employed in the reference pan.

This reference material exhibited a linear calorimetric trace with no thermal transitions in the temperature range of interest (-100° to 160°C.).

After inserting the test and reference specimens, the DSC cell was cooled with liquid nitrogen and subsequently heated to 160°C. at the desired programmed heating rate. The thermogram obtained from the first heating cycle was discarded, since coalescence of the sample occurred complicating the initial DSC trace. Subsequent heating-cooling cycles yielded smooth thermograms. Normally, 3 to 4 cycles were employed with each test specimen to insure reproducible results. Representative DSC thermograms are given in Figure 1.

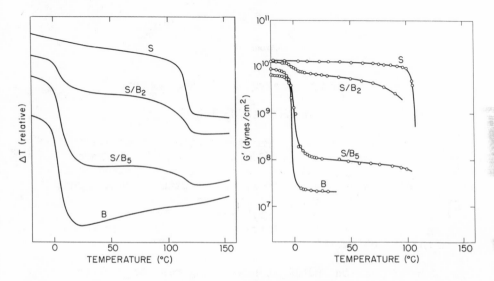

Figure 1. Typical DSC Thermograms; samples identified in Table I

Figure 2. Torsion Pendulum Shear Modulus vs. Temperature; samples identified in Table I

The graphical procedure used to assign the transition temperatures and calculate the relative magnitude of the heat capacity change (Δ) is illustrated in Figure 3. First, the base lines marked X were drawn tangent to the relatively flat plateau regions of the curve. Then the lines marked Y were drawn tangent to the steepest part of the curve in the "step" regions. A vertical line was then constructed midway between the intersections of line Y and base lines X. The vertical distance between the two base lines was employed as a measure of the relative heat capacity change (Δ) associated with the glass transition of the block segment. The glass transition temperature was chosen as the intersection of tangent line Y and the upper base line, a point near the onset of heat capacity change through the transition region.

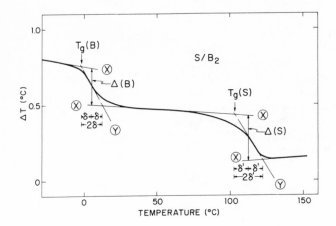

Figure 3. Method of Analyzing DSC Thermograms

RESULTS AND DISCUSSION

The completely amorphous polymers exhibited number average molecular weights ranging from 61,000 to 310,000[2]. Infrared and NMR analyses of the butadiene microstructure indicated that the monomer addition was primarily 1,2 (90%) with a small amount of trans-1,4 (10%). Similarly, the isoprene microstructure was mainly 3,4 and 1,2 with about 10% 1,4 adduct. The copolymer compositions were determined by infrared and NMR analyses which confirmed the stoichiometric results.

Ultracentrifugation and light scattering results verified that the copolymers had sharp block segments of high DP and that the samples had narrow molecular weight distributions. Specifically, the M_w/M_n values for the S, B and S/B$_2$ samples were 1.17, 1.08, and 1.22, respectively, indicating that the molecular weight distributions were narrow.

Light scattering measurements[2] were made on S/B$_2$ in a variety of solvents with systematically varying indices of refraction using the technique proposed by Stockmayer[3] and Benoit[4] for block copolymer characterization. The results summarized in Figure 4 show that the molecular weight is independent of the refractive index increments giving a flat linear plot, as expected for a block copolymer, whereas a random copolymer or a mixture would give a parabola with upward concavity[4]. These results are indicative of very little composition heterogeneity and we concluded, therefore, that the polymers are genuine block copolymers with insignificant quantities of homopolymer contamination.

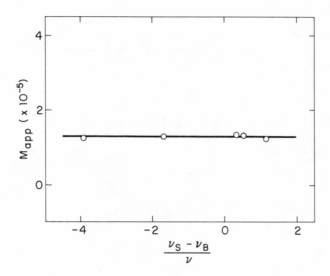

Figure 4. Apparent Molecular Weight vs. Refractive Index Increment Variable; ν_S is dn/dc of homopolystyrene, ν_B is the dn/dc of homopolybutadiene and ν is the dn/dc of the copolymer S/B2.

Differential scanning calorimetry (DSC) proved to be an excellent analytical technique for determining multiple glass transitions of this series of styrene-butadiene block copolymers. The existence of two distinct Tg regions (ca. 0°C. for 1,2-polybutadiene segments and ca. 100°C. for polystyrene segments) is clearly shown in the step-like nature of the DSC thermograms, as illustrated in Figure 1. These results are in agreement with our viscoelastic results, shown in Figure 2, where independent torsion pendulum measurements gave similar step-like behavior[1,2]. The DSC and torsion pendulum results illustrate the separate two phase nature of these block copolymers and demonstrate that the individual segment Tgs remain constant over the entire block copolymer composition range, in contrast with the behavior of typical random copolymers[5] where one composition-dependent Tg is observed. Table II summarizes the DSC results for the series of styrene-butadiene polymers.

DSC analyses of the homopolymer specimens, polystyrene (S), polybutadiene (B) and polyisoprene (I) were carried out as a function of heating rate. A typical example of the effect of heating rate is shown in Figure 5 for homopolybutadiene (B). We note that the relative heat capacity change (Λ) decreases as the heating rate decreases and extrapolates to a zero value at zero heating rate. Likewise, the measured glass transition temperature

decreases with decreasing heating rate, and with homopolybutadiene
extrapolates to a Tg value of -5°C. at an infinitely slow heating
rate.   Similar Tg data were obtained for S and I as a function of
heating rate and are listed in Table III.   The effects of heating
rate are in accord with published results[6].   We chose a 30°C./
minute heating rate in our DSC analyses for convenience and the
resulting clarity of the Tg inflections.

Table II

DSC Results for Styrene-Butadiene Polymers

Polymer Designation	Tg (B)* (°C.)	Tg (S)* (°C.)	Mole % S		
			DSC	NMR	Stoc.
S	–	111	100	100	100
S/B$_1$	-4	100	68	72	74
S/B$_2$	-1	104	55	55	59
S/B$_3$	-2	100	45	43	49
S/B$_4$	2	103	24	36	39
S/B$_5$	0	99	18	19	19
B	-1	–	0	0	0
B/S	-2	105	82	78	80

*DSC values at 30°C./min. heating rate

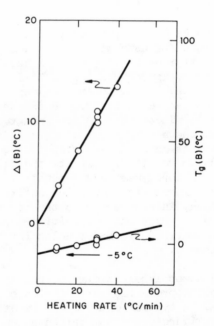

Figure 5. Effect of Heating Rate on Tg and Heat Capacity Change
for Homopolybutadiene

Table III

Homopolymer Data

Polymer Designation	Tg (°C.)*	DSC Relative Heat Capacity Change**	
		$(\frac{\Delta}{g})$ (°C./g x 10⁻¹)	$(\frac{\Delta}{g/m})$ (°C./mole x 10⁻³)
S	104	1.51	1.57
B	-5	2.90	1.57
I	11	2.30	1.56

*DSC values extrapolated to 0°C./min. heating rate
**DSC values at 30°C./min. heating rate

Gibbs and DiMarzio[7] report that the increase in heat capacity at Tg is due to increased configurational entropy and that the change in heat capacity per mole of repeat unit should be the same for all polymers which fit the same space lattice. Therefore, we expect that the heat capacity changes at Tg, per mole of repeat unit, would be the same for the three vinyl homopolymers employed in this study.

In our experiments, the molar heat capacity change (ΔC_p) is related to measured parameters by the following equation,

$$\Delta C_p = \frac{K \Delta}{g/m}$$

where K is an instrument constant, Δ is the relative heat capacity change as measured from our DSC thermograms, g is the sample weight and m is the molecular weight of a repeat unit.

The relative changes of heat capacity per unit weight of sample (Δ/g) and per mole of repeat unit

$$(\frac{\Delta}{g/m})$$

for the three homopolymers are listed in Table III. Within experimental error, the relative heat capacity changes per mole of repeat unit are equivalent for the three homopolymers. Therefore, from the equation above, it follows that the molar heat capacity change (ΔC_p) at Tg is the same for the three homopolymers.

These results thus permit us to determine the composition of the copolymers from the DSC thermograms, since the magnitudes of the Tg steps are proportional to the number of mole units of each component. By simply determining the ratio of the magnitudes of the two steps in an individual thermogram, i.e., $\Delta(B)/\Delta(S)$, the copolymer composition was easily calculated. The results, listed as mole % styrene, are given in Tables II and IV. The polymer compositions determined by DSC analyses agree with our previous NMR and stoichiometric results[2].

Table IV

Tg Trends for Polydiene Segments

Polymer Designation	Tg (°C.) Mechanical*	DSC**	Mole % S DSC	NMR
I	16	15	0	0
S/I	12	11	54	51
B/S/I	7	8	56	50
S/B/I	4	3	44	53
S/B$_3$	-2	-2	45	43
B	0-2	-1	0	0

*Torsion pendulum loss modulus max.
**Heating rate of 30°C./min.

In the series of styrene-butadiene block copolymers, the invariance of the individual segment transition temperature with composition is adequately demonstrated by the DSC and torsion pendulum results. These independent calorimetric and viscoelastic analyses illustrate the distinct two phase nature of the styrene-butadiene block copolymers and indicate the purity of the phase separations of the "hard" and "soft" segments.

However, with butadiene and isoprene block segments in the same copolymer chain, we find for the diene block segments a single Tg region about midway between the Tg regions of the respective butadiene and isoprene homopolymers. These DSC results are in accord with our earlier torsion pendulum results[2]. The DSC and torsion pendulum data for the two homopolymers I and B, and for the block polymers S/I, S/B/I, B/S/I and S/B$_3$ are summarized in Table IV. We interpret the results as relatively compatible phase blending of similar diene "soft" segments when incorporated in the same polymer chain, as opposed to the distinctly separate phase behavior observed with block copolymers of styrene ("hard" phase) and a single diene ("soft" phase).

REFERENCES

1. R. J. Angelo, R. M. Ikeda and M. L. Wallach, Paper presented at the Twelfth Canadian High Polymer Forum, St. Marguerite, Quebec, Canada, May, 1964.

2. R. J. Angelo, R. M. Ikeda and M. L. Wallach, Polymer, 6, 141 (1965).

3. W. H. Stockmayer, L. D. Moore, M. Fixman and B. N. Epstein, J. Polymer Sci., 16, 517 (1955).

4. W. Bushuk and H. Benoit, Can. J. Chem., 36, 1616 (1958); M. Leng and H. Benoit, J. Chim. Phys., 58, 479 (1961).

5. K. H. Illers, Kolloid Z., 190, 16 (1963).

6. S. Strella, J. Appl. Polymer Sci., 7, 569 (1963).

7. J. H. Gibbs and E. A. DiMarzio, J. Chem. Phys., 28, 373 (1958).

STUDIES ON DOMAIN FORMATION MECHANISM OF A-B TYPE BLOCK COPOLYMER FROM ITS SOLUTION[*]

T. INOUE, T. SOEN, T. HASHIMOTO[**], and H. KAWAI

Department of Polymer Chemistry, Faculty of
Engineering, Kyoto University, Kyoto, Japan

INTRODUCTION

Recent development in anionic polymerization technique have
made it possible to synthesize A-B-A as well as A-B type block
copolymer, and some characteristic mechanical properties of the
copolymers consisting of hard and soft block segments, such as
high-elasticity without crosslinking process, have attracted
attention (1). The dynamic mechanical properties, such as two
distinct peaks in the loss modulus-temperature relation and a
two step character of the dynamic modulus-temperature relation
including the high-elasticity, have been explained in terms of
a two-phase structure originating from the microphase separation
of the block segments into respective domains (1-12).

It has been suggested by Sadron et al. (13,14) and Vanzo
(15) that the two-phase structure of block copolymers in the
solid state is influenced by the configurations of the polymer
chains in the solutions at a critical concentration during the
solvent casting process. On the other hand, Riess et al. (16,17)
and Molau et al. (18-21) have carried out experimental studies on
the emulsifying effects of the block and graft copolymers, which
may restrain the phase separation of mixed system of the corres-

---

* Presented partly before the International Symposium on Macromolecular Chemistry, Tronto, Canada, September 5, 1968 and
partly before the Division of Polymer Chemistry, ACS New York
Meeting, New York City, September 8, 1969.

** Present address: Polymer Research Institute, University of
Massachusetts. Amherst, Mass. 01002, U.S.A.

ponding homopolymers into their microscopic domains and keeping
the system as microheterogeneous one.

In this paper, the domain formation mechanism of A-B type
block copolymer of styrene and isoprene from its solutions will
be, first, dicussed in terms of thermodynamic and molecular
parameters, i.e., the free energy of micelle formation at the
critical concentration during the solvent casting process (22).
Then, the domain structures of solvent cast films of several
compositions of three component system, A-B type block copolymer
of styrene and isoprene, polystyrene, and polyisoprene, will be
demonstrated, and the mechanism of domain formation will be
further discussed qualitatively in terms of thermodynamic parameters of phase equilibria in the four component system including
the solvent (23), taking account of the inherent nature of domain
formation of the A-B type block copolymer discussed in the former
part of this article.

EXPERIMENTAL PROCEDURES AND TEST SPECIMENS

A-B type block copolymers of styrene and isoprene, homo-
polystyrene, and homo-polyisoprene were synthesized by the
living polymerization technique initiated by n-BuLi in tetrahydrofuran at -78°C. Characterization of the polymers was
performed by ultracentrifugation, osmotic pressure, and ultra-
violet and infrared spectra as in a previous letter (24). The

Table 1. Characterization of Block Copolymers Synthesized

Block Copolymer	$M_n \times 10^{-4}$ [*]	Wt. fraction[**] of styrene sequence, %
70/30 Styrene-isoprene	104	73
50/50 Styrene-isoprene	70.1	49
40/60 Styrene-isoprene	53.8	42.5
20/80 Styrene-isoprene	27.8	18.5

[*] Measured by a high speed membrane osmometer in toluene solution at 37°C.

[**] Determined from UV absorption at 262 mμ in CCl_4 solution.

molecular weight of each block copolymers and its composition
in weight fraction thus determined are listed in Table 1.
The molecular weight of each homopolymer will be indicated
later in triangular diagrams showing the composition of the
binary or ternary system of the cast film.

Film specimens were, in general, cast from 5% solutions
(total polymer concentration) of mono, di or tri components of
the polymers in toluene by pouring onto glass plates and evapo-
lating the solvent gradually at 30°C., except for the cases of
examining the effects of initial polymer concentration of the
solutions and solvents themselves upon the domain structures.
The transparency of the cast films is indicated in terms the
circular symbols in the triangular diagrams; an open circle
designates transparent, a double open circle designates transparent but iridescent, a dot-open circle designates cloudy and
iridescent, and a dot designates opaque.

The domain structure of the cast film was observed by
light-microscopy and by electron-microscopy using OsO_4 fixation
technique (25). Thin sections, about 350 Å thick for electron--
microscopy and 5μ thick for light-microscopy, which were "stained
and fixed" with OsO_4 and then cut normal to the film surface,
were examined.

DOMAIN STRUCTURE OF A-B TYPE BLOCK COPOLYMER
OF STYRENE AND ISOPRENE FROM ITS SOLUTION

Change of Domain Structure with Fraction of A-B Sequence

For the specimens cast from 5% toluene solutions, the domain
formation of the two-phase structure originating from the microphase separation of the block segments was observed. As can be
seen in Fig. 1, the domain structure changed in a systematic
manner with increasing fraction of isoprene block sequence from
spheres of isoprene component dispersed in a matrix of styrene
component (Fig. 1-a), to zebra patterns of alternate strips of
each component (Figs. 1-b and 1-c), to spheres of styrene component (Fig. 1-d). It has been clarified stereographically that
the zebra pattern of alternate strips of each component is a
sectional view of an alternating arrangement of the two components,
i.e., each component separates into lamellae oriented parallel
to the film surface (24). The iridescent color effect, which had
been already pointed out by Vanzo and ascribed to the periodic
nature of the structure (15), was also observed, especially for
the 40/60 and 50/50 copolymers, in the solid state as well as
swollen states when the polymer concentration during solvent
casting became greater than a critical concentration of a little

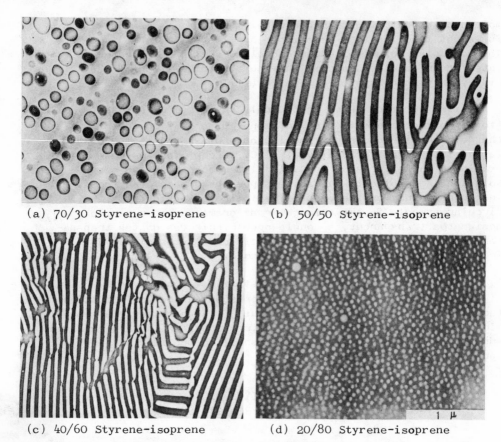

(a) 70/30 Styrene-isoprene   (b) 50/50 Styrene-isoprene

(c) 40/60 Styrene-isoprene   (d) 20/80 Styrene-isoprene

Fig. 1. Electron micrographs of ultrathin sections about 350 Å
thick, cut normal to the surface of films cast from 5% toluene
solutions of styrene-isoprene block copolymers varying in composition from 20 to 70% weight fraction of styrene sequences.

higher than 10%.

Change of Domain Structure with Casting Solvent

In order to investigate the effect of casting solvent on
the domain structure, the 40/60 block copolymer was cast from
solutions of about 2.5% concentration in various solvents at
room temperature. Electron micrographs are illustrated in Fig. 2.
The domain structure of the 40/60 block copolymer cast from
MEK (methyl ethyl ketone) (Fig. 2-a), cyclohexane (Fig. 2-b),
carbon tetrachloride (Fig. 2-c), n-hexane (Fig. 2-d), and isooctane (not shown but almost the same as Fig. 2-d), may be
classified, in general, as polystyrene fragments dispersed in

polyisoprene matrix in contrast to the alternate lamellar
arrangement observed for toluene irrespective the initial concentration of casting solution, which will be demonstrated in the
following section.

The shape and size of the polystyrene fragments are different from one solvent to another, probably reflecting different
solvating power for the block segments of the copolymers. It is
difficult, however, to demonstrate a systematic change of the
domain structure with solvation because of the lack of the exact
knowledge of this parameter, except for the qualitative representation in terms of the solubility parameters, such as Small has
proposed (26).

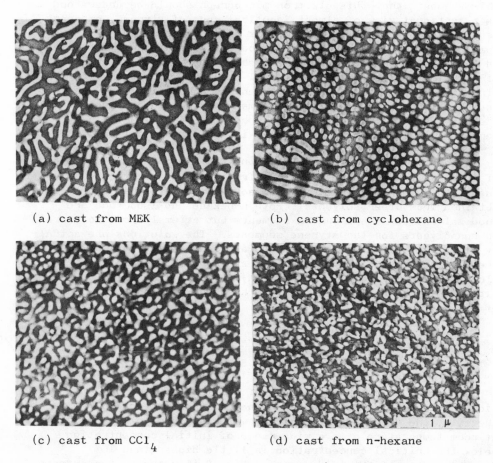

  (a) cast from MEK   (b) cast from cyclohexane

  (c) cast from CCl$_4$ (d) cast from n-hexane

Fig. 2. Electron micrographs of ultrathin sections about 350 Å
thick, cut normal to the surface of films cast from 2.5% solutions
of 40/60 styrene-isoprene block copolymer in various solvents.

The domain structures cast from MEK, carbon tetrachloride, and cyclohexane seem to be more irregular than the alternating lamellar structure obtained from toluene solution. Of these solvents, MEK may be classified as a good solvent for polystyrene but a relatively poor solvent for polyisoprene segments, carbon tetrachloride and cyclohexane are just the opposite, whereas toluene is a rather good solvent for both segments.

The electron micrograph of the specimen cast from cyclohexane (Fig. 2-b) reveals only two kinds of patterns of the unstained styrene component; i.e., circular patterns almost identical in diameter and parallel strips of various length and almost the same width as the diameter of the circles. These results are in contrast to the much irregular patterns of the other specimens. This suggests that this electron micrograph should be understood as a sectional view of almost identical rods of styrene component arranged almost parallel in a matrix of isoprene component. The domain structure observed here of rods of one component dispersed in a matrix of the other component should be considered as one of the simplest but most fundamental structure, together with the structures of spheres of one component in a matrix of the other and the alternating lamellar arragement of the two components. The existence of the rod structure had also been clearly demonstrated by Matsuo (27,28), as will be discussed later.

The structures cast from n-hexane and isooctane are composed of much smaller and more irregular fragments of the styrene comp- onent, which are dispersed in the isoprene matrix, than those obtained from other solvents. Both n-hexane and isooctane are good solvent for the isoprene segments but extemely poor solvents or nonsolvents for the styrene segments. The solutions are actual- ly milky white and must be regarded as pseudo-solutions in which the precipitated polystyrene chains are kept in suspension by the block segments of polyisoprene which are so well-solvated and expanded in the solvent as to give the above-mentioned structure of the matrix.

Effect of Initial Polymer Concentration of Solution upon Domain Structure

In order to check the effect of initial polymer concentra- tion in the casting solution upon the domain structure formed in the film, the 40/60 block copolymer was cast from toluene solution at room temperature over a wide range of initial concentration below the critical concentration (a little higher than 10%) for appearance of the iridescence. Fig. 3 illustrates two extreme results for a 0.8% solution (Fig. 3-a) and a 10% solution (Fig. 3-b), which give essentially the same structure, the alternating lamellar arrangement of both components, as that illustrated in

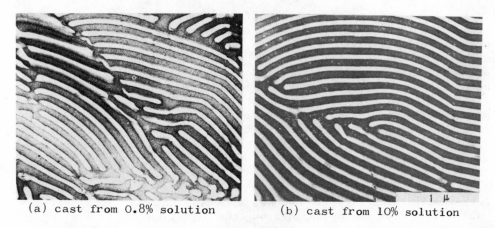

(a) cast from 0.8% solution   (b) cast from 10% solution

Fig. 3. Electron micrographs of ultrathin sections about 350 Å
thick, cut normal to the surface of films cast from 0.8% and
10% solutions of 40/60 styrene-isoprene block copolymer in toluene.

Fig. 1-c, obtained from a 5% solution. This suggests that the
domain structure is unaffected so long as the initial concentration is kept below the critical concentration.

Actually, iridescence appeared for all solutions tested
when evapolation during casting brought the concentration to
its critical value, a little higher than 10% in this system.
Thus it appears that the domain structure originates as an equilibrium phenomenon at the critical concentration irrespective of
the initial concentration, and carried over into the solid state.

Fundamental Domain Structure and Molecular Orientation

As has been shown in the previous section, the two-phase
microheterogeneous structures of the A-B type block copolymers
cast from their solutions may be represented by the following
three types of fundamental domain structures:
a) spheres of one component dispersed in a matrix of the other
component;
b) rods of one component, almost parallel to each other, dispersed in a matrix of the other component; and an
c) alternating lamellar arrangement of the two components.

When the A-B type block copolymer chains are organized
into these fundamental domain structures, all of the junction
points of the block segments must be arranged at the interface
between the two phases. The elements of the respective domain
structures and the molecular arrangements within the domains are
illustrated schematically in the upperhalf of Fig. 4.

By assuming that the density of each component within the
domain structure is identical with that of each homopolymer in
bulk, the effectively occupied area per chain block at the
interface may be evaluated from the average dimensions of the
domain structures and the average degrees of polymerization of
the block segments. The occupied area per chain block thus
evaluated for each type of domain structure are illustrated in
the lowerhalf of Fig. 4, together with the dimensions of each
structure evaluated from the respective electron micrographs as
average values (24). It is noted that the calculated areas are
all close to 1600 $Å^2$.

When the areas of 1600 $Å^2$ are compared with the dimensions
of the domain structure, it becomes apparent that the block chains
must be stretched and oriented along the direction perpendicular
to the interface; i.e., radial direction for the spherical and
rodlike structures and normal direction to the lamellar surface
for the alternating lamellar arrangement. This has been confirmed
experimentally by means of polarized infrared dichroism using
films of the 40/60 and 50/50 block copolymers cast from toluene
solutions, which give, as previously demonstrated, alternating
lamellar structures oriented parallel to the film surface. Some
parallel dichroisms of 889, 909 and 1493 cm^{-1} absorption bands
with respect to the direction normal to the film surface were

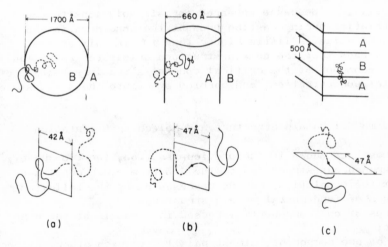

(a)   (b)   (c)

Fig. 4. Schematic representations of respective elements of three
types of fundamental domain structure and effectively occupied
area per chain in the interface between the two separate components: a) sphere of one component (isoprene) in a matrix of the
other component (styrene); b) rod of one component (styrene) in a
matrix of the other component (isoprene); and c) alternating lamellar arrangement of the two components (styrene and isoprene), which
correspond to Figs. 1-d, 2-b and 1-b, respectively.

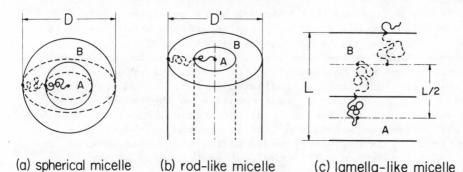

(a) spherical micelle   (b) rod-like micelle   (c) lamella-like micelle

Fig. 5. Schematic representations of three types of micelles
(component A dispersed in component B): (a) spherical micelle;
(b) rodlike micelle; (c) alternating lamellar micelle.

observed with polarized infrared radiation directed to the tilted
film surface at an angle of 30° from the film normal.

THERMODYNAMIC THEORY OF DOMAIN FORMATION OF A-B TYPE
BLOCK COPOLYMER FROM ITS SOLUTION (22)

As has been pointed out in the previous sections, there is
probably a critical concentration C^* at which each segment of the
block copolymers undergoes phase separation and aggregates into
characteristic molecular micelles, just as soap molecules do in
aqueous solution at a critical micelle concentration. The micelle
structures thus formed may be maintained as a whole at higher
concentrations until the solid structures are formed. The stability of the block copolymer micelles will be discussed in terms
of coalescence barrier similar to that of POO-emulsions, to be
discussed in a later section.

In this section, the manner of formation of the micelle will
be discussed in terms of an equilibrium at the critical concentration in order to present a mechanism for the formation of three
types of fundamental domain structure in terms of molecular and
thermodynamic parameters.

Let us introduce three types of micelles, namely, as illustrated in Fig. 5, spherical, rodlike, and alternating lamellar
micelles, which correspond to the three types of fundamental
domain structures that can form the domain structure at the critical concentration by the hexagonal closepacking (spherical and
rodlike micelles) or by piling up (lamellar micelles).

It may be assumed that when A-B block copolymer chains aggregate into such a micelle, phase separation occurs so as to form an interface and to arrange all junction points of the copolymer chains on the interfaces. The area of the interface per unit volume may be evaluated from dimensions, D, D' or L of the respective micelles, as indicated in Fig. 5, and the volume fraction, V_A, of A block segment relative to the total volume occupied by the copolymer chain at the critical micelle concentration, unless the volume fraction is changed by the micelle formation. V_A is defined by means of the volumes of spheres whose diameters are equal to the root mean squares of gyration of the chain blocks.

The Gibbs free energy G per unit volume for the formation of respective micelles may be given by:
for spherical micelle;

$$G_s = 6V_A^{2/3}(1/D)\Delta W + (3/2)kTN[(D^2/4)\{(1 - V_A^{1/3})^2/\sigma_B^2 n_B a_B^2$$
$$+ V_A^{2/3}/(\sigma_A^2 n_A a_A^2)\} - 2] \qquad (1-a)$$

for rodlike micelle;

$$G_r = 4V_A^{1/2}(1/D')\Delta W + (3/2)kTN[(D'^2/4)\{(1 - V_A^{1/2})^2/(\sigma_B^2 n_B a_B^2)$$
$$+ V_A/(\sigma_A^2 n_A a_A^2)\} - 2] \qquad (1-b)$$

for lamellar micelle;

$$G_\ell = 2(1/L)\Delta W + (3/2)kTN[(L^2/4)\{(1 - V_A)^2/(\sigma_B^2 n_B a_B^2)$$
$$+ V_A^2/(\sigma_A^2 n_A a_A^2)\} - 2] \qquad (1-c)$$

where ΔW is the interfacial contact energy between the two components per unit area (which is the contact energy difference between unlike segments in the sense of the quasi-chemical approach and may be repulsive for incompatible components), N is the number of copolymer chains per unit volume, σ_i^2 is a parameter characterizing the effect of solvent in terms of a ratio of the mean square of the end-to-end distance of the i-block at the critical micelle concentration to that of the chain with freely jointed segments, $\langle R_i^2 \rangle / \langle R_i^2 \rangle_f$, n_i is the degree of polymerization of the i-block sequence having monomer length a_i, k is Boltzman constant, and T is the casting temperature.

For the derivation of free energy of the micelle formation given by Eqs.(1), the reference state was taken as the virtual one with N chains of A-homopolymer, with $\langle R_A^2 \rangle$, and N chains of B-homopolymer, with $\langle R_B^2 \rangle$, in a state of complete phase separation into macroscopic domains. The entropy change per copolymer chain was approximated by simply assuming that the copolymer chain is composed of two Gaussian chains of A and B homopolymers, neglecting the effect of the junction between the A and B blocks to form the copolymer chain; i.e.,

$$\Delta S = \sum_{i=A,B} \Delta S_i = -(3/2)k \sum_{i=A,B} (r_{i,2}^2 - r_{i,1}^2)/(\sigma_i^2 n_i a_i^2) \qquad (2)$$

where $r_{i,1}$ and $r_{i,2}$ are the end-to-end distance of the i-block at the reference state and in the micelle, respectively. The end-to-end distance, $r_{i,2}$, in the micelle formation was also approximated so that the junction point is fixed at the interface of respective micelles, while the ends of each block chain are fixed at the particular points in the micelles, as illustrated in Fig. 5. The approximation of $r_{i,2}$ is very crude, giving a too high density at the center of the spherical or rodlike micelle and too low density at the outer surface of the micelle. An opposite evaluation of the configurational entropy has been proposed by Meier (29) using the diffusion equation to represent chain configuration so as to make the density within the domain constant in the solid state.

Fig. 6 shows the free energy curve derived from Eqs.(1) as a function of micelle size. The curve, which has a free energy trough as a result of the summation of the first term (hyperbolic decrease with mecelle size) and the second term (2nd power increase

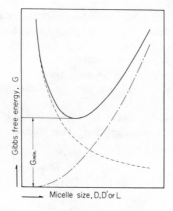

Fig. 6. Qualitative illustration of the relationship between the free energy of micelle formation G, and the size of the micelle, D, D', or L.

with micelle size) on the righthand side of Eqs.(1), indicates
that the micelle should have the size giving the minimum free
energy G_{min}. In addition, it may be noted that the minimum free
energy is retained as positive in contrast to a mechanical mixture
of two incompatible components, for which the free energy change
on mixing must be zero as a result of the complete phase separation.
In other words, the microphase separation to form micelles of
particular size is the most stable for A-B block copolymer, in
contrast to complete phase separation into macroscopic domains,
which is the stable state for a mixture of the incompatible A- and
B-homopolymers.

More quantitative evaluation of the conditions for the
minimum free energy can be made by differentiating Eqs.(1) with
respect to the micelle size as follows:

$$D_{G,min} = f_s(\phi, \sigma, n)\gamma, \text{ where } f_s \text{ is given by} \tag{3-a}$$

$$f_s \equiv [8V_A^{2/3}/\{(1 - V_A^{1/3})^2/(\sigma_B^2 n_B a_B^2) + V_A^{2/3}/(\sigma_A^2 n_A a_A^2)\}]^{1/3}$$

$$D'_{G,min} = f_r(\phi, \sigma, n)\gamma, \text{ where } f_r \text{ is given by} \tag{3-b}$$

$$f_r \equiv [(16/3)V_A^{1/2}/\{(1 - V_A^{1/2})^2/\sigma_B^2 n_B a_B^2 + V_A/(\sigma_A^2 n_A a_A^2)\}]^{1/3}$$

$$L_{G,min} = f_1(\phi, \sigma, n)\gamma, \text{ where } f_1 \text{ is given by} \tag{3-c}$$

$$f_1 \equiv [(8/3)/\{(1 - V_A)^2/(\sigma_B^2 n_B a_B^2) + V_A^2/\sigma_A n_A a_A^2)\}]^{1/3}$$

where $\gamma^3 \equiv \Delta W/kTN$, and $\sigma \equiv \sigma_B/\sigma_A$.

Equation (3) predict that the size of the micelle increases
as the cube root of the degree of polymerization $n = n_A + n_B$,
with increasing ΔW, and with decreasing casting temperature T,
provided that the temperature dependence of σ and ΔW are negligibly small. The ΔW dependence means that the more incompatible
are the A and B components, the larger are the micelles.

By substituting Eqs.(3) into the respective Eqs.(1), the
minimum values of the free energy for the formation of the
respective types of the micelles, $G_{i,min}$ are given by:

$$G_{s,min} = \left\{ 3V_A^{2/3}f^{-1} - (1/\gamma^2) \right\} 3kTN\gamma^2 \tag{4-a}$$

$$G_{r,min} = \left\{ 2V_A^{1/2}f_r^{-1} - (1/\gamma^2) \right\} 3kTN\gamma^2 \tag{4-b}$$

$$G_{1,min} = \left\{ f^{-1} - (1/\gamma^2) \right\} 3kTN\gamma^2 \tag{4-c}$$

Comparison of $G_{i,min}$ among the three types of micelle can
be made by examining the first term of the righthand side of
Eqs(4). Figure 7 shows the above relative values of minimum free
energies as a function of weight fraction ϕ_A of A-blocks, where
the numerical values, $\sigma_B/\sigma_A = 1$, $a_A = a_B = a$, and $m_A/m_B = 104/68$,
are assumed for comparing the results with styrene-isoprene block
copolymers cast from toluene solution. Since the formation of
micelles having the lowest relative value of the minimum free
energy at a given weight fraction ϕ_A should predominate (as
illustrated by thicker lines in the figure) at the equilibrium
state at C^* and T, it appears that the type of micelle, and hence
the resulting domain structure of the bulk specimen, changes in
order from spherical to rodlike to alternating lamellar with
increasing ϕ_A. These expectations agree at least qualitatively
with the experimental results, shown in Fig. 1. The values ϕ_1
and ϕ_2 indicated in the figure give the upper (lower) limit of
the weight fraction of the blocks for formation of each particular
micelle type.

The free energy minimum for micelle formation may be the
characteristic thermodynamic feature of A-B block copolymer. It
is treated in this paper as an entropic contribution to the free
energy of micelle formation (the second term on the lefthand
side of Eqs.(1)) which opposes the phase separation due to the
incompatibility of the two components (the first term on the
lefthand side of Eqs. (1)), i.e., the block segments are expected
to orient to a considerable extent along the direction perpendi-

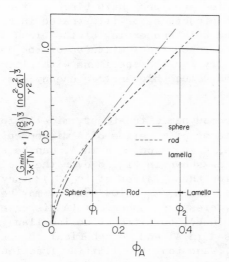

Fig. 7. Relationship between the relative minimum free energy of
micelle formation (A dispersed in B) and the weight fraction ϕ_A
of A-block (with assumed values, $\sigma_B/\sigma_A = 1$, $a_A = a_B = a$).

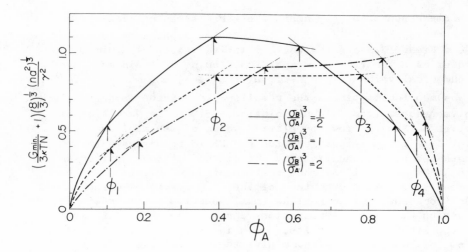

Fig. 8. Effect of solvent upon the relationship between the
relative minimum free energy of micelle formation and the weight
fraction ϕ_A of A-blocks. The change of mode of micelle formation
from A dispersed in B (when ϕ_A is less than ϕ_2) to B dispersed
in A (when ϕ_A is larger than ϕ_3) is indicated.

cular to the interface between the two components. This particular
orientation-aggregation of the block chains at the interface
must be reflected in the rather small ratio of the dimension of
the effectively occupied area per chain block at the interfaces
to that of the domain structure, as illustrated in Fig. 4, and
must further cause the bulk properties of the block copolymer
to differ from those of mechanical mixtures of the corresponding
homopolymers, especially for the specimens with alternating lamellar structure which have much higher free energy, as shown in
Fig. 7, than the others.

In the above thermodynamic treatment, micelle formation was
mainly discussed in terms of micelles of A dispersed in component
B. The opposite case, dispersion of B in A, can be similarly
treated. The result is shown in Fig. 8 over the whole range of
ϕ_A; plots are given for these values of $(\sigma_B/\sigma_A)^3$ varying from
1/2 to 2, $a_A = a_B = a$, and $m_A/m_B = 104/68$. It may be seen that
the type of micelle varies from spherical (A dispersed in B), to
rodlike (A dispersed in B), to alternating lamellar (A and B),
to rodlike (B dispersed in A) and to spherical (B dispersed in
A) with increasing ϕ_A, and that the limiting fraction ϕ_i for
each type of micelle shifts toward lower values with decreasing
relative solvation power, σ_B/σ_A.

This dependence of the type of micelle on ϕ_A agrees, at least qualitatively, with the changes in the domain structures shown in Fig. 1, except for the rodlike structures which were unfortunately unrealized because of the lack of the copolymers containing adequate weight fractions. Quite recently, the existence of the rodlike structures as well as the above change of domain structure as a function of ϕ_A have been more completely demonstrated by Matsuo et al. (28). They formed the same morphological effects in A-B, A-B-A and A-B-A-B block copolymers of styrene and butadiene.

The change of domain structure of the 40/60 block copolymer from the alternating lamellar structure (Fig. 1-c) to the styrene rods in an isoprene matrix (Fig. 2-b) is represented in Fig. 8 by changing the relative solvation power $(\sigma_B/\sigma_A)^3$ from unity to 2 at $\phi_A \cong 0.4$; i.e., the change of solvent from toluene (a rather good solvent for both blocks) to cyclohexane (a good solvent for the isoprene block but a rather poor solvent for the styrene block) may be represented in terms of the change in the relative solvation power from almost unity to larger than unity.

The independence of the limiting fraction ϕ_i with respect to temperature (insensitivity of the type of micelle to casting temperature) has been checked; however, the dependence of the size of micelle could not be found experimentally (22). Quite recently, the dependence of the size of micelle with respect to the total degree of polymerization n was discussed experimentally by Bradford and Vanzo (29).

It may be concluded in this part that one can account for the five types of fundamental domain structures and their size, obtained by casting A-B type block copolymers from solutions, in terms of thermodynamic and molecular parameters, such as the incompatibility ΔW between the block segments, the relative solvation power σ of the solvent for each blocks, the total chain length n of the copolymer, and the molecular composition ϕ_A.

DOMAIN STRUCTURES OF TERNARY POLYMER SYSTEM OF A-B TYPE BLOCK COPOLYMER OF STYRENE AND ISOPRENE, POLYSTYRENE AND POLYISOPRENE CAST FROM SOLUTIONS (23)

Experimental Results

Figs. 9 through 11 show the electron micrographs of the ternary or binary system, together with the triangular diagrams indicating the composition of the systems and the transparency of the film specimens.

All specimens in Fig. 9 are mixtures in which the ratio of styrene to isoprene components is maintained at 73/27 which is similar to styrene/isoprene ratio in the 70/30 block copolymer. The molecular weights of the homopolymers are also similar to those of the corresponding copolymer blocks. The method for forming the domain structure of the 70/30 block copolymer from toluene solution, i.e., to form spheres of isoprene component dispersed in a matrix of styrene component (Fig. 9-a), is still maintained for the ternary mixtures along the isopleth, as seen in Figs. 9-b through 9-f. The size distribution of the domain structure is considerably broadened as the fraction of the copolymer decreases. It appears that each homopolymer added is solubilized into respective block domains composed of like components.

For the ternary mixtures along the isopleth corresponding to the composition of 40/60 block copolymer, the original domain structure with the alternating lamellar arrangement, as seen in Fig. 10-a, is maintained only in a large fraction of the copolymer and is gradually changed, as illustrated in Figs. 10-b through 10-e with decreasing of the fraction of the copolymer, to a structure of very irregular fragments of styrene component dispersed in a matrix of isoprene component. For the binary mixture of the 40/60 block copolymer with homo-polystyrene or homo-polyisoprene, where the total composition of styrene or isoprene sequences is not kept constant, the domain structure changes systematically with increasing of the fraction of styrene or isoprene sequences; i.e., for the former case, the structure changes from the alternate lamellar arrangements, Fig. 10-a, to more separated lamellar arrangement, Figs. 10-f and 10-g, due to thickening of styrene lamella, while for the latter case, the structure changes from the alternate lamellar arrangement to styrene rods, Fig. 10-h, to styrene sphere, Fig. 10-i, both dispersed in a matrix of isoprene component. For every specimen in Fig. 10, where the molecular weights of the homopolymers are again similar to those of the corresponding blocks of the copolymer, the solubilization of the homopolymers into the respective block domains take place. But the domain structure of alternate lamellar arrangement for the 40/60 block copolymer is not maintained if the copolymer fraction along the isopleth drops below around 50%.

For the binary mixtures of hsM-series in Fig. 11, where the 20/80 block copolymer is mixed with homo-polystyrene of relatively higher molecular weight than that of corresponding block of the copolymer, it is noticeable that a phase separation into the homo-polystyrene phase and the block copolymer phase occurs without any disturbance of original domain structure of the block copolymer. This is obvious in a comparison of Fig. 11-b' an electron micrograph of the block copolymer phase (matrix phase in Fig. 11-b) with Fig. 11-a, the original domain structure of the

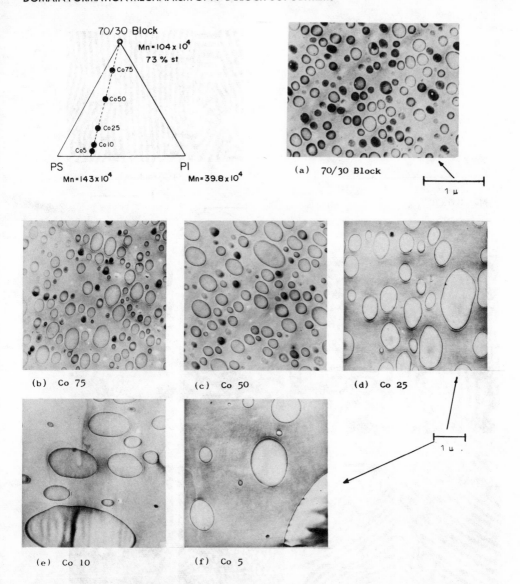

Fig. 9. Triangular diagram representing the composition in weight
fractions of ternary mixtures and electron micrographs of ultrathin sections of films cast from 5% toluene solutions and stained
by OsO_4. (Evidently, the matrix domain in Figs. 9-c through 9-e,
is darker than the spherical domain. But, when one re-stains the
ultrathin sections with OsO_4, the spherical domain becomes quite
dark. This suggests that OsO_4 diffusing through the matrix domain
deposits on the surface of the spherical domains of the isoprene
component and reduces to metallic osmium preventing further diffusion into the spherical domain.)

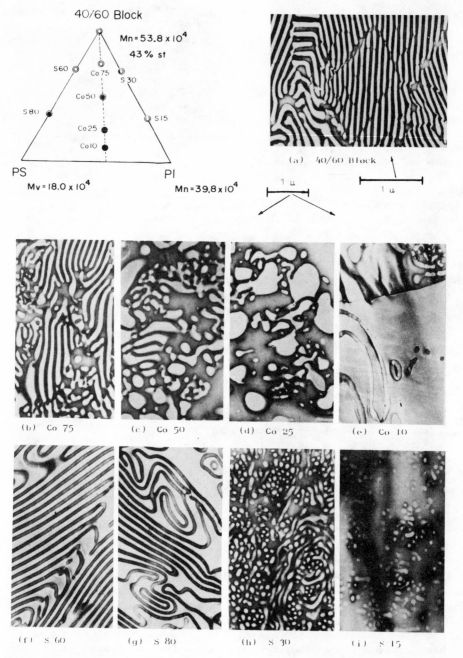

Fig. 10. Triangular diagram representing the composition in
weight fractions of ternary or binary mixtures and electron micrographs of ultrathin sections of films cast from 5% toluene solutions
and stained by OsO_4.

20/80 block copolymer. Large ellipsoids of the homo-polystyrene
component are dispersed in a matrix of the block copolymer component which forms an inherent domain structure of spheres of
styrene blocks in a matrix of isoprene blocks. The same phenomena
are observed in a mixture of the block copolymer with a homo-polystyrene of much higher molecular weight than the hsM-series, as
illustrated in Fig. 11-f. In this case, the interface between
the homo-polystyrene phase and the block copolymer phase is
demonstrated clearly in Fig. 11-f'. On the other hand, when the
block copolymer is mixed with a homo-polystyrene of relatively low
molecular weight, that is, lower than that of the corresponding
block of the copolymer, the homo-polystyrene is solubilized into
the domain of styrene block of the copolymer, as seen in Fig. 11-g.
This is in contrast to the above cases involving high molecular
weight polystyrene but is similar to the cases discussed in connection with Figs. 9 and 10. For the binary mixtures of 20/80
block copolymer with a homo-polyisoprene, whose molecular weight
is of the same order as that of the corresponding block of the
copolymer, the homo-polyisoprene solubilized into the matrix domain
of isoprene block of the copolymer. This increases the separation
of the spheres of styrene blocks, as seen in Fig. 11-h in comparison with that in Fig. 11-a.

The above results indicate that when the molecular weight
of homopolymer added is of the same order or less than that of
the corresponding block of the copolymer, the solubilization of
the homopolymer into the domain of the corresponding block takes
place. In this case, the two incompatible homopolymers can be
blended well with each other by adding the corresponding block
copolymer. That is the block copolymer behaves just like an
emulsifier restraining the phase separation of the homopolymers
into their macroscopic domains. But when the molecular weight of
the homopolymer added is much higher than that of the corresponding block, the copolymer can no longer act as the emulsifier.
Namely, the block copolymer behaves as if it were incompatible
with the corresponding homopolymers and loses the tendency to
incorporate the sequences of like homopolymers.

The inherent manner of domain formation of the block copolymer
from its solution was discussed in the previous part in terms of
the thermodynamic and molecular parameters. Results presented
here suggest that when the above condition for the solubilization
of homopolymers into the block domains of the copolymer is satisfied and the fraction of the copolymer is kept relatively large,
the same manner of domain formation may be maintained even for
the ternary and binary systems. Namely, the formation of five
types of fundamental domain structures, A spheres in B matrix,
A rods in B matrix, alternating lamellar arrangement, B rods
in A matrix, and B spheres in A matrix, may be achieved, depend-

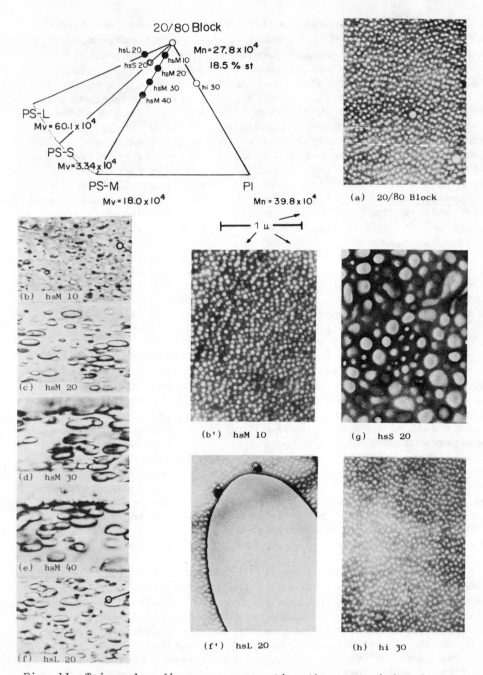

Fig. 11. Triangular diagram representing the composition in weight fractions of binary mixtures and electron and ordinary micrographs of ultrathin and thin sections of films cast from 5% toluene solutions and stained by OsO_4.

ing upon the total ratio of A sequences to B sequences in the
system. This is demonstrated schematically in Fig. 12, using the
triangular phase diagram, in which the chain line indicates the
isopleth keeping the total sequence ratio identical with that of
the A-B type block copolymer. Thus, for the diagrams from Figs.
12-a through 12-c the composition of the A-B type block copolymer
changes from A rich copolymer to rather B rich copolymer. In each
diagram the area 1 corresponds to the domain structure of A spheres
dispersed in B matrix, the area 2 to A rods in B matrix, the area
3 to alternating lamellar arrangement, the area 4 to rods in A
matrix, and the area 5 to B spheres in A matrix. The above domain
formations may be achieved stably only outside of the hatched area.
The results of ternary system in Fig. 9 correspond to the area 5
in Fig. 12-a when the component A is styrene. The results of
binary systems in Fig. 10 along the sides of 40/60 Block-PS
and 40/60 Block-PI correspond to the side AB-A within the area
3 and the side AB-B in Fig. 12-c, respectively, when the component
A is isoprene.

    Judging from the relationship between the transparency of
cast films represented by the circular symbols and their domain
structures, it may be said that the transparency depends on the
size and distribution of the domain structure and that iridescent
phenomena are due to the periodic nature of the structure, both
in relation to the wave length of visible rays.

Domain Formation Mechanism of the Ternary or Binary System

    Molau et al.(18-20) demonstrated a mechanism for the domain
formation of rubber particles during polymerization of solution

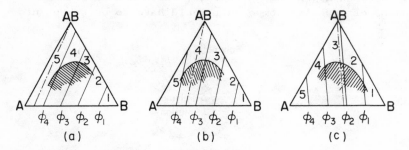

(a)   (b)   (c)

Fig. 12. Triangular phase diagrams representing the mode of domain
structures formed from solutions of the three polymers in a
common solvent, where the chain line indicates isopleth, and the
area 1 corresponds to the domain structure of A spheres dispersed
in B matrix, the area 2 to A rods in B matrix, the area 3 to the
alternate lamellar arrangement, the area 4 to B rods in A matrix,
and the area 5 to B spheres in A matrix.

of rubber in vinyl monomers. They showed that a polymeric oil-
in-oil emulsion, produced after a phase inversion at the beginning
of the polymerization, is transformed into a solid structure as
the polymerization proceeds. Assuming that the residual monomer
at the stage of POO-emulsion is considered as a casting solvent,
and that the graft copolymer of the rubber and vinyl polymer
produced during the polymerization is an A-B type block copolymer,
the polymerization process may be treated as similar to the solvent
casting process in our experiments. Consider the process of casting from a dilute solution, the three polymer solutes, A-B type
block copolymer, A-homopolymer, and B-homopolymer, in a common
solvent to form the ternary solid. Here a POO-emulsion thus
formed is maintained as a whole at higher concentration until the
solid structure is achieved. The block copolymer chains are
squeezed out on the interface of the POO-emulsion (see Fig. 13)
as a result of the complete immiscibility (31,32) of the unlike
sequences at sufficiently high polymer concentration.

The formation of a micro-heterogeneous domain structure of
two incompatible components by the solvent casting of the three
polymer solutes, as demonstrated in Figs. 9 through 11, suggests
the emulsifying effect of the A-B type block copolymer. This
effect must now be discussed in terms of the POO-emulsion state.
The uniformity of the domain size as well as the regularity of
the domain shape in the cast films suggest that a high degree of
stability of the POO-emulsion is achieved by the coalescence
barrier which may prevent demixing of the system. This is indicated
by the data in the figures, especially when the fraction of the
block copolymer in the system is relatively large. In addition
to Molau's discussion concerning sources of the coalescence
barrier in POO-emulsions (18-21), the concept of the entropic
repulsion between approaching droplets, which are covered by
block chains of opposite component, would have to be taken into

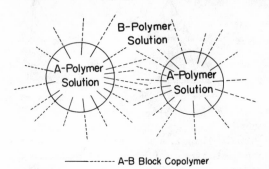

Fig. 13. Polymeric oil-in-oil emulsion of A droplets-in-B solution
type, stabilized by the A-B type block copolymers accumulated at
the interface between the spherical droplets and the matrix.

account in the way proposed by Clayfield (33) and Meier (34).
The workers consider the repulsive force between particles, which
absorb polymer chains on each surface, in terms of decrease in
configurational entropy of the chains due to the loss of possible
configulations as the space available to the chains is reduced
between approaching particles. A broad size distribution with
decreasing fraction of the A–B type block copolymer added, is
observed along the isopleth in Fig. 9. Such behavior can be
explained in terms of a random aggregation of the droplets of the
POO-emulsions having low coalescence barriers due to the lack of
copolymer chains being supplied on the unit surface area of the
droplets. The irregularity of domain shape, which is observed
along the isopleth in Fig. 10, may be also explained in terms of
the lack of sufficient copolymer chains because of the relative
large interface area of the alternating lamellar arrangement
compared to the spherical domain structure.

The next problem to be discussed here is the criterion for
the solubilization of the four component system, A–B type block
copolymer, A-polymer, B-polymer, and solvent. For the four
component system, there may also be a critical concentration of
total polymer, C^*, at which the POO-emulsion is formed as a result
of microphase separation. When C^* of the four component system
is at C_1^* in Fig. 14, the solubilization has to occur in the
following fashion. If C^* is outside of the binodal surface of the
four component phase diagram, the solution is homogeneous at C^*
and the three polymer solutes cooperate together for the micelle
formation which results in the POO-emulsion stabilized by the
copolymer. On the contrary, if the C^* is at C_2^* in the diagram of
Fig. 14, i.e., inside of the binodal surface, the phase separation
into the copolymer solution and the homopolymer solutions occurs
before C^* is reached during the casting process. Consequently,
the block copolymer forms an inherent domain structure undisturbed
by the homopolymers, resulting in a separation into macroscopic
phases of the copolymer and the homopolymers. Therefore, the
problem is one of relating C^* to the binodal surface.

A theoretical derivation of C^* may be performed by develop-
ing Meier's formulation (30), but this seems unlikely to give an
analytical solution. Experimental values of C^*, such as those
obtained by Vanzo as around 9% for a system of ethylbenzene and
a block copolymer of styrene and butadiene (15), may be utilized.

Derivation of the binodal surface of the four component
system may be very complicated. Even for the ternary system, it
is difficult to get an analytical representation of the binodal
curve. However, in considering the three components without one
of the homopolymers, the generality for the problem is not lost,
and a plait point of the system composed of A–B type block copolymer, B-polymer and solvent may approximated as (23), by using

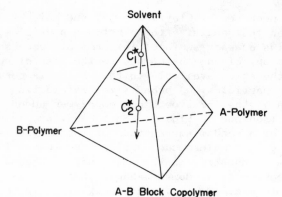

Fig. 14. Schematic representation of the binodal surface in the phase diagram of a four component system, where C^* is the critical micelle concentration of the system.

the Flory-Huggins theory (35) and the dilution approximation (32),

$$C_c = 1 - (v_s)_{crit.} = (1 + 1/\sqrt{r})^2/(2 \chi \psi_A) \qquad (5)$$

where C_c is the total polymer concentration at a plait point, and $(v_s)_{crit.}$ is the volume fraction of the solvent in the ternary system at the critical concentration. Equation (5) may predict the compatibility of the A-B type block copolymer with the B, ratio of the degrees of polymerization, r, the composition of the block copolymer, ψ_A, and the copolymer-homopolymer interaction parameter, χ .

Consequently, one may judge roughly whether the solubilization occurs or not, from the following condition;

$$2 \chi \psi_A c^* \gtrless (1 + 1/\sqrt{r})^2 \qquad (6)$$

ACKNOWLEDGEMENTS

The authors are deeply indebted to Dr. K. Kato and Mr. M. Nishimura, Central Research Laboratory, Toyo Rayon Co., Ltd., for kindly arranging for preparation of the electron micrographs, and to Prof. M. Kurata, Institute for Chemical Research, Kyoto University, for his valuable comments and discussions. A part of this work was supported by sientific research grant from the Japan Synthetic Rubber Co., Ltd.

REFERENCES

1) R.E. Gunningham and M.R. Treiber, J. Appl. Polymer Sci., $\underline{12}$, 23 (1968)
2) A.A. Rembaum, F.R. Ells, R.C. Morrow, and A.V. Tobolsky, J. Polymer Sci., $\underline{61}$, 155 (1962)
3) A.V. Tobolsky and A. Rembaum, J. Appl. Polymer Sci., $\underline{8}$, 307 (1964)
4) M. Bear, J. Polymer Sci., A, $\underline{2}$, 417 (1964)
5) R.J. Angelo, R.M. Ikeda, and M.L. Wallach, Polymer, $\underline{4}$, 141 (1965)
6) S.L. Cooper and A.V. Tobolsky, J. Appl. Polymer Sci., $\underline{11}$, 1361 (1967)
7) G. Kraus, C.W. Childers, and T. Gruver, J. Appl. Polymer Sci., $\underline{11}$, 1581 (1967)
8) C.W. Childers and J.T. Gruver, J. Appl. Polymer Sci., $\underline{11}$, 2121 (1967)
9) C.W. Childers and G. Kraus, Rubber Chem. Technol., $\underline{40}$, 1153 (1967)
10) H. Hendus, K.H. Illers, and E. Ropte, Kolloid-Z., u. Z. Polymere, $\underline{216-217}$, 110 (1967)
11) E. Fischer and J.F. Henderson, Rubber Chem. Technol., $\underline{40}$, 1373 (1967)
12) N.H. Canter, J. Polymer Sci., A-2, $\underline{6}$, 155 (1968)
13) C. Sadron, Angew. Chem., $\underline{75}$, 472 (1963)
14) B. Gallot, B. Mayer, and C. Sadron, Compt. Rend. Acad. Sci., Paris, $\underline{263C}$, 42 (1966)
15) E. Vanzo, J. Polymer Sci., A-1, $\underline{4}$, 1727 (1966)
16) G. Riess, J. Kohler, C. Tournut, and A. Banderet, Macromol. Chem., $\underline{101}$, 58 (1967)
17) J. Kohler, G. Riess, and A. Banderet, European Polymer J. $\underline{4}$, 173 (1968)
18) G.E. Molau, J. Polymer Sci., $\underline{A-3}$, 1267 (1965)
19) G.E. Molau, J. Polymer Sci., $\underline{A-3}$, 4235 (1965)
20) G.E. Molau and H. Keskkula, J. Polymer Sci., A-1, $\underline{4}$, 1595 (1966)
21) G.E. Molau and W.M. Wittbrodt, Macromolecules, $\underline{1}$, 260 (1968)
22) T. Inoue, T. Soen, T. Hashimoto, and H. Kawai, J. Polymer Sci., A-2, $\underline{7}$, in press
23) T. Inoue, T. Soen, T. Hashimoto, and H. Kawai, Macromolecules, in press.
24) T. Inoue, T. Soen, H. Kawai, M. Fukatsu, and M. Kurata, J. Polymer Sci., B, $\underline{6}$, 75 (1968)
25) K. Kato, Polymer Eng. Sci., $\underline{7}$, 38 (1967)
26) P.A. Small, J. Appl. Chem., $\underline{3}$, 71 (1953)
27) M. Matsuo, S. Sagae, and Y. Jyo, J. Electron Microscopy, Japan, $\underline{17}$, 309 (1968)
28) M. Matsuo, S. Sagae, and H. Asai, Polymer, $\underline{10}$, 79 (1969)
29) E.B. Bradford and E. Vanzo, J. Polymer Sci., A-1, $\underline{6}$, 1661 (1968)

30) D.J. Meier, J. Polymer Sci., C, <u>26</u>, 81 (1969)

31) A. Dobry and F. Boyer-Kawenoki, J. Polymer Sci., <u>2</u>, 90 (1947)

32) R.S. Scott, J. Chem. Phys., <u>17</u>, 279 (1949)

33) E.J. Clayfield and E.C. Lumb, J. Colloid Interface Sci.,
    <u>22</u>, 269 (1966)

34) D.J. Meier, J. Phys. Chem., <u>71</u>, 1861 (1967)

35) P.J. Flory, "Principles of Polymer Chemistry", Ithaca, N.Y.
    (1953)

COLLOIDAL AND MORPHOLOGICAL BEHAVIOR OF BLOCK AND GRAFT COPOLYMERS

Gunther E. Molau

The Dow Chemical Company

Midland, Michigan 48640

INTRODUCTION

Block and graft copolymers can be visualized as homopolymers which are connected by chemical bonds. Molecules of block and graft copolymers are thus molecules of a higher order. They are more complex than molecules of either homopolymers or random copolymers, and they are more difficult to synthesize in a well-defined form, but they have also many properties which cannot be found in homopolymers or random copolymers. If the different sequences of the homopolymeric substructures are long enough, the properties of both of the corresponding homopolymers can often be found simultaneously in block or graft copolymers. On the other hand, if the homopolymeric substructures are short, block and graft copolymers can behave more like random copolymers and can have properties intermediate between those of the corresponding homopolymers. Block and graft copolymers also can have properties found neither in homopolymers nor in random copolymers. These properties are a result of the composite nature of the block or graft copolymer molecules which permit not only interactions between the polymer chains and the environment, e.g. the solvent when in a solution, but also interactions between the different homopolymeric subchains themselves.

These types of interactions give block and graft copolymers their colloidal and interfacial properties. In solution, interactions between the homopolymeric subchains and the solvent stabilize organic sols and organic latices (1-3), and interactions between different homopolymer chains and block or graft copolymer subchains stabilize polymeric oil-in-oil emulsions (4,5). In the solid state, polymer-polymer interactions lead to new morphologies

and to novel mechanical properties which have recently found com-
mercial use in the thermoplastic rubbers marketed by the Shell
Chemical Company under the trade names THERMOLASTIC[®] and
KRATON[®] (6-8).

The earlier work on block and graft copolymers had centered
on the design of materials with novel or improved mechanical properties and an extensive literature (9-11) on the preparation and
characterization of block and graft copolymers developed. In
terms of commercial success, the return has been meager in comparison to many other areas of endeavor. It is interesting to
note that even the most recent monograph (11), in agreement with
preceding surveys, lists only six instances of commercial applications of block or graft copolymer systems. Most of the encountered
difficulties are of a preparative nature. The older methods of
preparation are often inefficient and give block or graft copolymers in low yields, accompanied by large amounts of homopolymers.
Fractionation of these mixtures, even in analytical quantities, is
difficult, often impossible, and very few graft copolymers have
been fully characterized (10), a situation which appears to persist
at the time of this writing.

Anionic methods of polymerization (12,13) and organometallic
methods of synthesis (14-16) are being more and more frequently
applied to block and graft copolymer synthesis. Although these
new methods are not without unsolved problems, mainly because they
require starting materials of extreme purity, they have the poten-
tial of overcoming the two major deficiencies of previous methods
of preparation: low yield and undefined structure. Many block
copolymers of well-defined structure have been prepared recently
by anionic polymerization, particularly block copolymers of styrene
with methylmethacrylate, butadiene or isoprene. The availability
of these new block copolymers has shifted the interest of many researchers from the mere design of new materials to studies of the
fine structure of the copolymers, their solution behavior and
their colloidal properties. An extension of these studies to well-
defined graft copolymers can probably be expected in the foresee-
able future, since organometallic methods of synthesis make these
copolymers accessible.

The present paper is a review of the published work on the
colloidal and morphological behavior of block and graft copolymers.
In order to avoid tiresome repetition of the words "block and graft
copolymers" or "block or graft copolymers," the abbreviation "BG
copolymers" will be used for both expressions.

COLLOIDAL PROPERTIES BASED ON POLYMER-SOLVENT INTERACTIONS

Since BG copolymers consist of homopolymeric substructures, they can have dual solubility in some solvents or solvent mixtures. In the extreme case, the solubilities of the parent homopolymers may differ so widely that the corresponding BG copolymers can be used as emulsifiers for oil-water systems, much like the conventional soaps or low molecular weight nonionic surfactants. Well-known examples are a group of ethylene oxide-propylene oxide block copolymers which are marketed as surfactants under the trade names PLURONICS® and TETRONICS® (17). Block copolymers (18-21) and graft copolymers (22,23) of styrene and ethylene oxide have also been used as emulsifying agents in oil-water systems. These BG copolymers differ from conventional, low-molecular weight surfactants only in their molecular weight; their mode of action is probably not essentially different.

The first colloidal properties of graft copolymers were discovered and recognized by Merrett (1) in the course of studies on the grafting of natural rubber with styrene or methylmethacrylate. Merrett fractionated the primary reaction mixture, which consisted of free rubber, grafted rubber, and a vinyl polymer, by dissolving the polymer mixture in benzene (a good solvent for all constituents) and precipitating one homopolymer selectively. The specific nonsolvents were methanol, which precipitates rubber, and petroleum ether, which precipitates polystyrene or polymethylmethacrylate, respectively. In both cases, a stable sol was formed as an intermediate. Later, Merrett called these organic sols "organic latices" (2,3), because they resemble a latex in every respect, except that the water is replaced by an organic phase as the dispersion medium.

A number of authors (13, 24-29) have reported the formation of colloidal sols of a similar nature. In most cases, the sols were obtained in the course of separations of mixtures of BG copolymers from the corresponding homopolymers by fractional precipitation. Sol formation occurred always at the point of precipitation of one of the homopolymers, i.e. when the solvent/nonsolvent mixture became a selective solvent for only one of the parent homopolymers. The stability of many of these nonaqueous colloids is fantastically high. In organic sols stabilized with grafted rubbers, repeated cycles of freezing in liquid nitrogen followed by thawing do not lead to flocculation (31). Coagulation can often be achieved only by high speed centrifugation for extended periods of time.

The stability of these nonaqueous sols can be understood on the basis of polymer-solvent interactions. Merrett (2,3) and other authors (28,29) showed that at the point of precipitation

of one of the parent homopolymers, the corresponding subchains of
the BG copolymer are also precipitated. The collapsed chains form
the nuclei of micelles or "organic latex" (3) particles, depending
on the system. If pure BG copolymers are used, the term micelles
(2,29) may be more appropriate. In analogy to soap micelles,
these polymeric micelles are aggregates of a number of BG copolymer molecules. A homopolymer which corresponds to the collapsed
portion of the molecules is occluded in the micelles, quite analogous to "solubilization" of oils in soap micelles in water, and
"organic latex" particles are formed. In either case, the colloidal stability of the particles has been attributed to the dual
solubility of the BG copolymer molecules: at the point at which
part of the BG copolymer molecule is collapsed, as a result of
selective precipitation by the specific nonsolvent, the other portion of the molecule is still soluble and holds the entire particle, i.e. the micelle or the "organic latex" particle, in
colloidal dispersion (2,3,28,29). The possibility cannot be excluded that stabilization by polymer-polymer interaction, as
discussed later on, also plays a role in selective solvent systems.
In these systems, polymer-polymer interactions may stabilize in
addition to the polymer-solvent interactions discussed above.

The architecture of the micelles or "organic latex" particles
(Fig. 1) is determined by the solvent-nonsolvent system. The
polymer chains or subchains, which are precipitated first, form
the micellar cores. These cores are surrounded by shells of the
still soluble subchains of the BG copolymer molecules. Soluble
homopolymer molecules, which may also be present, probably are

POLY-A   POLY-B
PRECIPITATED   DISSOLVED

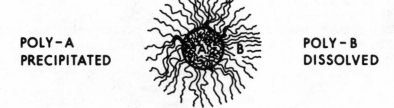

Fig. 1. Formation of micelles or latex particles in
         selective solvent systems.

distributed over the entire solvent phase. Bresler and co-workers
(28) have suggested different micelle structures for various types
of block copolymers. The core-shell architecture is preserved in
the solid polymers which can be prepared from the sols either by
flocculation or by solvent evaporation. Thus, BG copolymers can
be obtained in two modifications by proper choice of the precipi-
tants (Fig. 2). This effect is particularly spectacular in systems in which one of the involved chain-types is elastomeric while
the other chain-type belongs to a rigid polymer. In such a case
the two modifications appear as a "hard form" and a "soft form,"
because the overall properties of the composite polymer are determined mainly by the material which forms the micellar shells.

The formation of hard and soft modifications of natural
rubber grafted with rigid polymers has been demonstrated and explained by Merrett (30). Bresler and co-workers (28) obtained
similar modifications with styrene-isoprene block copolymers. The
explanation of the formation of modifications was based on the
partial collapsing of the block or graft copolymer molecules at a
certain solvent/nonsolvent composition which was demonstrated by
measurements of the intrinsic viscosity as a function of solvent
composition (2,28,29,32). More recently, the morphology of hard
and soft modification of styrene-butadiene block copolymers has
been studied by electron microscopy and the earlier conclusions
have been confirmed by direct visual evidence (33).

At very low concentrations of BG copolymers, the micelles are
apparently monomolecular, because light scattering indicated no

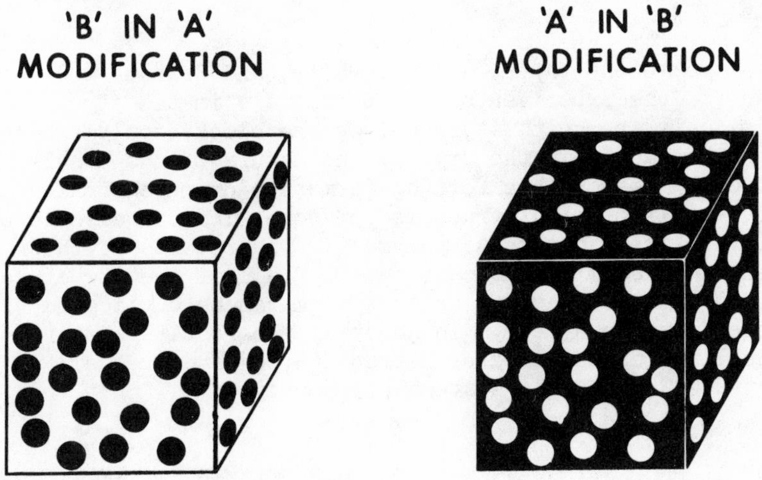

Fig. 2. Formation of two modifications from selective
         solvent systems.

increase in molecular weight as a consequence of micelle formation (29). At higher concentrations, aggregation of molecules
and formation of polymolecular micelles seemed to occur (Fig. 1).
Other factors influencing the number of molecules per micelle
were the temperature and the quality of the solvent (32). The
generality of the results obtained by Gallot and co-workers (29,
32) has been confirmed by studies of micelle formation of other
graft copolymers (34,35).

Studies of the solution behavior of styrene-ethylene oxide
block copolymers in selective solvents by Skoulios and co-workers
(18-21) have led to the discovery of polymeric liquid crystals.
In dilute solutions, the styrene-ethylene oxide block copolymers
formed monomolecular micelles in the same way as the BG copolymers
described above. At somewhat higher concentrations, aggregation
to polymolecular micelles occurred, but as the concentration was
increased further, highly organized, mesomorphic structures were
formed which could be detected by x-ray analysis or by polarized
light microscopy. Depending on the nature of the solvent and on
concentration and temperature, the molecules aggregated in cylindrical or in lamellar mesomorphic forms. At some concentrations,
both forms, the leaflets and the cylinders, could coexist.

Mechanical agitation or heat destroys the structure of the
mesomorphic gels. However, their structure can be permanently
fixed by using a vinyl monomer as a selective solvent and solidi-
fying the entire system by polymerization. The new composite
polymers accessible by this method have been described by Sadron
(37-39). He proposed the name "organized polymers" for this class
of materials.

In a more recent paper, Gallot, Mayer, and Sadron (40) repeated many of the experiments done previously in the styrene-
ethylene oxide system with styrene-isoprene block copolymers in
order to investigate, if the crystalline nature of the ethylene
oxide is solely responsible for the formation of mesomorphic
structures, or if ordered structures of some kind can also be obtained with entirely amorphous polymers. The results of this
more recent work seem to indicate that not all of the structures
found in the styrene-ethylene oxide system can be duplicated with
the styrene-isoprene system. In general, the results seem to
follow the lines established by Merrett (2,30) and Bresler (28)
who had studied comparable systems previously.

COLLOIDAL PROPERTIES BASED ON POLYMER-POLYMER INTERACTIONS

Polymer-polymer interactions within BG copolymer molecules
or between copolymer and homopolymer molecules lead to a spectrum

of colloidal properties as manifold as the colloidal properties
found in solution in selective solvents, where they are a result
of polymer-solvent interactions, i.e. a result of the nonuniform
solubility of the molecules. Moreover, polymer-polymer interactions not only determine the architecture and the properties of
BG copolymer molecules in solution, but also the morphology in the
solid state.

The nature of the polymer-polymer interactions within BG copolymer molecules or between these molecules and homopolymer constituents can be visualized by inspecting the behavior of mixtures
of homopolymers, either in solution or in the solid state. With
very few exceptions, mixtures of polymers are heterogeneous, because polymers of different chemical structure are usually incompatible. The generality of this phenomenon has first been
demonstrated by Dobry and Boyer-Kawenoki (41) who studied a large
variety of mixtures of homopolymers. Scott (42) has presented a
theoretical treatment of phase separation of polymers, and Flory
(43) added theoretical considerations which make clear that incompatibility of different polymers is expected to be the rule,
and compatibility is a rare exception. Stockmayer and co-workers
(44) pointed out that interactions, which are in their net effect
repulsive, must be considered even between different units in
random copolymers, and Molau (45) has shown that such interactions
are strong enough to cause phase separation even within a series
of random copolymers of identical monomers, but of different composition.

If phase separation occurs between unidentical chains of homopolymers, an intramolecular phase separation can be expected to
occur between unidentical subchains of BG copolymers. Without
question, incompatibility of chemically different chain ends of BG
copolymers must be expected in the solid state. In solution, even
in dilute solution, intramolecular phase separation should occur
because introduction of a solvent into mixtures of homopolymers
only dilutes the effect of incompatibility, but does not eliminate
it (41). The objection can be raised that intramolecular phase
separation should not occur in dilute solutions of BG copolymers
in a nonselective solvent, because it is known that two-phase
systems, polymer A - polymer B - solvent, become homogeneous at
high dilution (41). However, if phase separation within a molecule is considered, the concentrations of the various polymer
subchains have to be considered with respect to the volume occupied
by this one molecule. This means that the concentration of any of
the polymer subchains is much higher within the space occupied by
the molecule than with respect to the solution as a whole.

Repulsive polymer-polymer interactions and an intramolecular
phase separation within molecules of BG copolymers have been

postulated by a number of authors (5,13,24,35, 46-52), and experimental studies with regard to this problem have appeared in the
recent literature. Most of the experimental work has been investigations of the dilute solution properties of styrene-methyl-
methacrylate block copolymers. Burnett, Meares, and Paton (47)
and Urwin and Stearne (48,50) found that the Flory dilute solution
parameters θ, $\varkappa_2$, and ψ_2 of the block copolymers have higher values
than those of the corresponding homopolymers. Although the number
of samples studied was not sufficient to characterize the entire
curves, it appeared that plots of these parameters as functions of
the copolymer composition have maxima near the center of the composition scale. Plots of the configurational parameters as functions of copolymer composition indicated an increase of the average
dimensions of the block copolymer molecules over those of the
parent homopolymers of equal molecular weight. Both research
groups interpret their results using the following hypothesis:
Interactions of a repulsive nature exist between the chemically
different subchains of block and graft copolymer molecules and
these interactions lead to intramolecular phase separation and to
expansion of the molecules, particularly in good solvents where
the number of repulsive contacts reaches a maximum (50).

Further studies of styrene-methylmetnacrylate copolymers have
been reported by Krause (49), and Inagaki and Miyamoto (51,52) who
agree with the above-cited authors on the postulate of repulsive
interactions and intramolecular phase separation. However, the
low values of the intrinsic viscosity found by Sonja Krause seem
to contradict the idea of an increased expansion of block copolymer
molecules in solution. This apparent controversy may have been
solved by the data of Inagaki and Miyamoto (52) who found that the
intrinsic viscosity of the block copolymer is more strongly molecular weight dependent than that of the homopolymers. The depend-
ency is such that the block copolymers have a lower intrinsic
viscosity than the homopolymers in the lower molecular weight
range, but have a higher intrinsic viscosity than the homopolymers
in the higher molecular weight range.

A model of a dissolved BG copolymer molecule, which assumes
that the chemically different subchains occupy different domains
of the space filled by the entire molecule and the attached solvent molecules, would explain the experimentally observed phenomena discussed above and in the following sections of this review.
The model would also be consistent with the established incompatibility of different polymers on a macroscopic scale. The space
occupied by a BG copolymer molecule could have a shape similar to
a dumbbell, or the molecule could form a spherical micelle (5)
such that one subchain is coiled in the center of the micelle and
the other subchain is wrapped around this core, thus forming the
micellar shell. Similar spatial arrangements can be visualized

for multiblock copolymers or for graft copolymers with many side chains. Association of several BG copolymer molecules to form polymolecular micelles in the same basic spatial arrangements is possible.

Since the polymer chains are in continuous motion, the time-average mass distribution over the micellar space must be considered when intramolecular phase separation is discussed. Attempting to visualize various possible time-average mass distributions, one is reminded of the distribution of electrical charge in electron orbitals. For illustrative purposes, a dumbbell-shaped mass distribution space can be compared to a p-orbital (Fig. 3), and a spherical mass distribution is somewhat analogous to an s-orbital. A quantitative measure of intramolecular phase separation would result from a comparison of the time-average mass distribution of the subchains in a block copolymer to the corresponding distribution of hypothetical subchains in a homopolymer molecule of equal total length, where the hypothetical subchains are obtained by arbitrarily designating a portion of the homopolymer chain as A and another portion as B. If a true intramolecular phase separation exists in a block copolymer, the mass distribution of its chain ends would have to be substantially different from the mass distribution of the arbitrarily marked hypothetical subchains of a homopolymer of equal total chain length, i.e. the distance, d, between the centers of mass of the two portions, A and B, of the total chain would have to be significantly larger for a block copolymer $(A \neq B)$ than for a homopolymer $(A = B)$ (See Fig. 3).

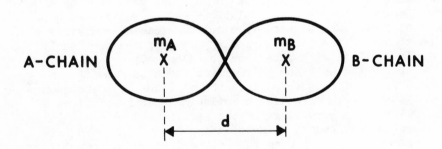

m_A, m_B = CENTERS OF MASS

Fig. 3.   Dumbbell model of the space occupied by an AB block copolymer chain.

Although measurements of thermodynamic solution parameters
have contributed a great deal to the understanding of the solution
behavior of BG copolymers and have provided strong evidence for
the two-phase model described above, the final proof or disproof
of the model would have to come from morphological studies. The
anticipated size range of the expected micelles (5) falls within
the range amenable to electron microscopic investigation, but a
twofold difficulty has to be overcome: (1) the BG copolymer solution must be fixed, i.e. solidified, in order to permit the preparation of specimens suitable for electron microscopy, and (2)
the phases must be made visible by some means of introducing contrast.

A new method for introducing contrast into polymer specimens
containing unsaturation recently has been introduced by Kato (53).
In his original paper, Kato had stained the rubber phase in
rubber-modified polymers with osmium tetroxide. This treatment
introduces contrast in the electron beam and hardens the rubber,
thus permitting slicing of ultrathin samples. Although this method
has been used by biochemists, it was apparently unknown to most
polymer chemists. (Staining with osmium tetroxide has been used
by Andrews (81) in studies on the crystallization of natural
rubber.) To researchers working in the area of rubber-modified
polymers, Kato's method has been a major breakthrough.

Kato's staining method has been utilized in a recent morphological study (33) of block copolymers in solution. Samples amenable to electron-microscopic investigation have been prepared
using two simple artifices: styrene-butadiene block copolymers
have been employed because the butadiene subchains can be stained
with osmium tetroxide, and styrene monomer has been used as a nonselective solvent, because the solutions can be solidified by
polymerization of the solvent. The solid specimens contained
micelles of one of the two spherical types proposed previously
(5), namely micelles with butadiene cores and styrene shells.
When free polybutadiene homopolymer was added to some of the
solutions, it was incorporated into the micelles, a phenomenon
analogous to the solubilization of oils in aqueous solutions of
soap micelles. The morphology of the specimens depended on the
polymerization conditions in the fixation step: spherical micelles were obtained only when the polymerization was carried out
with agitation. Specimens prepared without agitation contained
layers or cylinders rather than spheres. The sensitivity of agitation has its parallel in the finding of Skoulios and co-workers
(18-21, 36) that mesomorphic gels obtained with styrene-ethylene
oxide copolymers in selective solvents were also sensitive to
agitation. The possibility cannot be ruled out that the observed
spherical micelles are entirely an artifact of the sample preparation technique and that the undisturbed solutions contained

lamellae or cylinders similar to those observed by Skoulios'
group under different conditions. On the other hand, a spherical
morphology has also been obtained without any polymerization,
namely in films cast from solutions of block copolymers and homopolymers in benzene (33).

Since butadiene chains and styrene chains are incompatible,
the formation of polystyrene in a polymerizing solution of
styrene-butadiene block copolymer in styrene can be expected to
enhance the formation of micelles. Therefore, the occurrence of
micelles or lamellae in the fixed samples does not provide con-
clusive evidence for the existence of such structures prior to
fixation by polymerization. However, even if the micellar or
lamellar structures in the fixed samples owe their existence
entirely to the presence of polystyrene, their formation would
still be a result of polymer-polymer interactions. The effect
of polymer-solvent interactions, as discussed in the preceding
section of this review, must be excluded, because such interaction forces act on all portions of the molecules alike, since
styrene is a truly nonselective solvent for styrene-butadiene
copolymers and their parent homopolymers.

Quite generally, polymer-polymer interactions can be studied
in solution without undue interference of polymer-solvent interactions, if the polymer components and the solvent are chosen
such that the polymers have similar chemical structures, polari-
ties, and solubility parameters (i.e. the solvent is equally good
for all polymer chains in the system). In addition to formation
of micelles or other types of oriented structures, BG copolymers
have other colloidal properties which are entirely a result of
polymer-polymer interactions, because they are found in truly
nonselective solvents.

Another example illustrating the colloidal properties of BG
copolymers in nonselective solvents is the emulsifying effect of
these copolymers in "polymeric oil-in-oil emulsions"(4,5,54,55).
Polymeric oil-in-oil emulsions are prepared by dissolving two
different polymers and the corresponding BG copolymer in a mutual
solvent. As a result of the incompatibility of chemically different polymers (41), two phases are formed. These phases are
immiscible oils, i.e. immiscible solutions of polymers in non-
polar solvents. Another route to polymeric oil-in-oil emulsions
is the polymerization of a solution of one polymer in a monomer
of a different type. In this case, both the second homopolymer
and a graft copolymer (the oil-in-oil emulsifier) are formed in
situ. The latter method of preparing polymeric oil-in-oil emulsions has considerable commercial significance, because it plays
an important role in the formation of certain rubber-modified
polymers (54). In the present context, polymeric oil-in-oil

emulsions are of interest, because the mechanism of their stabilization (5) is a good illustration of the effect of polymer-
polymer interactions.

    In order to stabilize an emulsion by establishing a barrier
to the coalescence of droplets, an emulsifier must accumulate in
the emulsion interface (Fig. 4). In conventional oil-water
emulsions, surfactants accumulate in the interface simply because
of their dual solubility in oil and water. In polymeric oil-in-
oil emulsions, a dual solubility in the conventional sense, i.e.
with respect to the solvent, cannot exist because the solvent is
the same in both phases. The two phases are immiscible not because solvents are incompatible, but because the polymeric solutes
are incompatible. In the ideal case, the solvent is nonselective,
i.e. equally good for all components of the system, homopolymer
chains or subchains of the BG copolymer emulsifiers alike, and
the phases have about the same polarity and the same solubility
parameters. The problem is then: How and why does the emulsifier
accumulate in the interface if not by dual solubility in the conventional sense, i.e. by interactions between the polymer chains
and the solvent? As hinted, above, the answer is that polymer-
polymer interactions are responsible for this accumulation. Thus,
the same phenomenon which causes polymer incompatibility, i.e.
which is responsible for the separation into immiscible oil
phases, is also the cause of the emulsifying power of BG copolymers. Only in the emulsion interface are all repulsive polymer-
polymer interactions balanced, and each set of subchains of the
block or graft copolymer is in an environment with which it is
compatible. A distinction between an accumulation in the interface due to repulsion and an accumulation due to attraction may

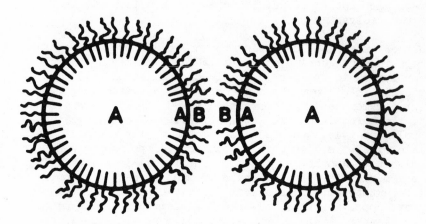

Fig. 4. Model of a polymeric oil-in-oil emulsion
         The block or graft copolymer (AB) accumulates
         in the emulsion interface.

seem either sophisticated or even far-fetched, but it is important
for understanding some of the phenomena observed in these systems.
The most typical phenomenon is the effect of the total molecular
weights of BG copolymers and the lengths of the individual subchains on the stability of the oil-in-oil emulsions. Roughly
speaking, below a total molecular weight of 100,000, the emulsifying power of block copolymers (and, apparently, graft copolymers) becomes insignificant (31). As the molecular weight increases, the emulsifying power increases and apparently levels
off at very high molecular weights. It is also important that
the individual homopolymeric subsequences of the block or graft
copolymers be fairly long if maximum emulsifying power is to be
achieved. Polymer incompatibility and the related repulsive
polymer-polymer interactions are entirely molecular weight phenomena (41-43) and have no parallel in low molecular weight compounds. Therefore, small molecules can have no emulsifying effect
at all in polymeric oil-in-oil emulsions. In contrast to the
decisive effect of the molecular weight in polymeric oil-in-oil
emulsions, small molecules as well as polymers can be used as nonionic surfactants in oil-water emulsions, because the emulsifying
power depends in oil-water systems on solubility effects alone
and not on polymer-polymer incompatibility.

POLYMER-POLYMER INTERACTIONS IN THE SOLID STATE

Although polymer-polymer interactions in the solid state and
their effect on the morphology of BG copolymers are not principally different from the interactions found in solution, studies
on solid samples are discussed in a separate section, because
substantial literature has evolved on the subject of the morphology and physical properties of solid BG copolymers. Most of
this literature has appeared within the last two years. Almost
all of the recent investigations deal with block copolymers of
styrene with either butadiene or isoprene. There are two reasons
for this preference: (1) Block copolymers with elastomeric subchains are commercially important. (2) Butadiene or isoprene
portions of block copolymer molecules can be stained selectively
with osmium tetroxide and are thus amenable to electron-microscopic
observation.

The first experimental results indicating that solid block
copolymers are two-phase systems came from studies of physical
properties rather than from morphological investigations.
Henderson, Grundig, and Fischer (56) measured the stress birefringence of styrene-isoprene block copolymers and came to the
conclusion that the styrene portions occupy different domains
than the isoprene portions of the molecules. Angelo, Ikeda, and
Wallach (57) studied transitions in block copolymers of styrene

either with isoprene or butadiene using a torsion pendulum. They found two glass transition temperatures, the values of which coincided with the values of the free homopolymers. The same two transitions were observed when physical blends of the homopolymers were prepared in the proportions given by the compositions of the block copolymers studied. Based on these data, the authors concluded that block copolymers are two-phase systems.

A new trend in research on block and graft copolymers is the increasing emphasis placed on morphological studies. Investigations of this nature have greatly increased the understanding of composite polymer systems, because they have not only revealed details about the morphology of block and graft copolymers and have confirmed the presence of two phases in these systems, but they have also shed new light on the problem of polymer and copolymer incompatibility.

Recently, Riess and co-workers (58-60) have reported studies of ternary mixtures of block copolymers (styrene with either isoprene or methylmethacrylate) with the corresponding homopolymers. The morphology of the samples has been investigated by phase microscopy and by electron microscopy. Maps showing areas of compatible as well as incompatible compositions have been obtained by drawing triangular diagrams and entering the compositions of the mixtures and the morphological observations. The diagrams show three regions: a region of turbid films which contained obviously two phases, an intermediate region of slightly hazy films, and a region of clear films which were either homogeneous or had inhomogeneities too small to scatter visible light. Investigation with the phase microscope confirmed the presence of two phases in the turbid films, but electron microscopy using the osmium tetroxide staining method revealed a two-phase structure also in most of the samples which appeared clear to the naked eye. The effect of the molecular weights and the compositions of the block copolymers on the morphology has been investigated, and the morphological results were related to measurements of the impact resistance of the samples.

In the last two to three years, several research groups (61-66) have reported morphological studies of block copolymers of styrene with either butadiene or isoprene, using solid samples which were not diluted with a solvent or with homopolymers. The block copolymers were prepared by anionic polymerization methods which assure well-defined compositions and narrow molecular weight distributions, if "killing" impurities in the monomers and solvents are meticulously avoided during the polymerization. Most block copolymers were either of the AB type (61,63,65,66) or of the ABA type (62, 64-65). Although principal differences in the morphology of these two groups have not been established conclusively, a

substantial difference in mechanical properties seems to exist.
The electron-microscopic investigations were based on two techniques: the preparation of surface replicas (61,63), and the
preparation of ultrathin sections, either parallel or normal to
the surface, which were stained with osmium tetroxide (62, 64-
66).

Using a replica technique, Vanzo (61) and Bradford and Vanzo
(63) studied the surface morphology of styrene-butadiene block
copolymers of the AB type and found patterns of layers or ripples
on the surface of their samples. Preparing cross-sections of the
same samples and staining them with osmium tetroxide, they found
patterns of black and white stripes in the interior of the
samples. The width of the stripes and their arrangment were
identical with those of the ripples found on the surface. The
patterns of stripes or ripples look like human fingerprints.
Since osmium tetroxide stains only the butadiene portions, the
photomicrographs of the cross-sections provide direct evidence
of the two-phase nature of the block copolymers. The stripes or
ripples represent an orientation of the styrene sequences and
the butadiene sequences in the form of layers which seem to extend throughout the samples. Evidence that the observed patterns
represent layers rather than another type of geometry has been
obtained by laying cross-sections in varying directions through
the samples: the patterns were essentially independent of the
direction of the cuts. The width of the stripes may vary with
different types of sample preparation, e.g. they were dependent
on solvent and concentration in the casting of films, but within
a series of samples of varying molecular weights, the spacing of
the layers increased with increasing molecular weight such that
the width of the stripes is proportional to the square root of
the molecular weights of the segments. The spacing of the layers
was in good agreement with the calculated root-mean-square end-
to-end distance of the segments.

The morphology of styrene-butadiene-styrene three-block copolymers has been studied by Hendus, Illers, and Ropte (62) and
by Beecher, Marker, Bradford, and Aggarwal (64). Cross-sections
of the samples were stained with osmium tetroxide. The morphology depended on the history of the samples: molding or extrusion
introduces strains which are visible in the electron photomicrographs. This observation has been made also in the case of
styrene-butadiene two-block copolymers (31). Strain-free samples
can best be prepared by dissolving the polymers and casting films.
In a series of samples with varying styrene-butadiene ratio,
Hendus et al. (62) found spherical particles of butadiene sequences embedded in a matrix of styrene sequences in samples of
block copolymers with a high styrene content. In samples with
low styrene content, they found the opposite morphology, namely

particles of styrene sequences embedded in a matrix of butadiene
sequences. In the intermediate composition range, patterns of
stripes were observed which resemble the patterns of stripes seen
by Bradford and Vanzo (63) in two-block copolymers of styrene and
butadiene. A variation of the morphology from spheres to stripes
with changing composition was observed by Inoue and co-workers
(66) in two-block copolymers of styrene and isoprene. Unfortunately, the exact geometry of the structures which appear as
stripes in the two-dimensional representation given by the photomicrographs of Hendus et al., Beecher et al., and Inoue et al.
is uncertain because the published pictures represent cross-
sections in only one direction.

Beecher et al. (64) propose a "cubic structure" for their
samples which is an arrangement of spherical particles in a cubic
lattice, similar to the crystal lattice of sodium chloride, but
the spherical polymer particles are interconnected by channels of
polymer, thus forming a three-dimensional network. A similar
cubic structure has been proposed by Gallot et al. (40) for
styrene-isoprene block copolymers partially precipitated from
selective solvents.

After comparing the influence of block copolymer composition
on morphology observed by Hendus et al. (62) and by Inoue et al.
(66) with the formation of micelles of block copolymers embedded
in homopolymer matrices reported by Molau (33), it appears that
the main factor determining whether spherical particles or layers
and rod-like structures are formed is the volume ratio of the two
phases, i.e. the phase of all styrene sequences and the phase of
all butadiene sequences. When a system contains homopolymers as
well as block copolymers, it seems to be immaterial whether polymer chains belong to block copolymer or homopolymer molecules.
Only the total volumes of the phases appear to determine the overall morphology. Thus, the same block copolymers which had a
layered structure in the undiluted state, as shown by Bradford
and Vanzo (63), formed spherical particles of butadiene sequences
when they were embedded in a matrix of polystyrene homopolymer
(33). This comparison shows that spherical particles can be
formed by two routes, either by increasing the amount of one sequence in the block copolymer or by adding the corresponding
homopolymer. The effect of the phase volume ratio on the overall
morphology has a parallel in polymeric oil-in-oil emulsions (4,54)
which also consist of two polymeric phases and where a change from
one type spherical morphology to the opposite type occurs as the
volume ratio of the two phases is changed. In the intermediate
range, the phase invert and patterns of stripes and layers are
observed quite analogously to the phase inversion seen in the
picture series of Hendus et al. (62) and Inoue et al. (66) where
the inversion is a result of the change in composition. It is

noteworthy that the dimensions of the stripes or particles in
this picture series are almost unaffected by the change in composition; all samples had about the same molecular weight. Although Hendus et al. (62) did not characterize their samples by
molecular weight determination, the widths of the stripes correlate
with the K-values, which are given instead of molecular weights.
As in the case of two-block copolymers (63), the width of the
stripes increases with increasing molecular weight. The dimensions
obtained directly from the electron photomicrographs were in close
agreement with values obtained by low-angle x-ray scattering. In
general, the x-ray values were about 40 Å higher than the microscopic values (62).

The morphology and the physical properties of styrene-butadiene
block copolymers of varying composition and varying number and
arrangement of the blocks has been studied recently by Matsuo and
co-workers (65). The following sequence arrangements have been
covered: SB, BSB, SBS, and SBSB. Specimens for electron microscopy were prepared from both compression-molded sheets and films
cast from solution. The preparation of specimens from solution
was the preferred procedure, because it minimizes the strain in
the samples. The sections were taken normal as well as parallel
to the surface of the specimens, and the ultrathin slices were
stained with osmium tetroxide in order to introduce contrast.

The SB and BSB block copolymers studied by Matsuo and co-
workers had intermediate compositions similar to the SB block
copolymers studied by Bradford and Vanzo (63), and in both papers,
sections were taken parallel as well as normal to the sample surfaces. In the photomicrographs of Bradford's samples, patterns
of black stripes are observed which are essentially invariant
under a change of the direction of the cut, indicating a layered
structure of the samples. However, the photomicrographs of
Matsuo's block copolymers showed black stripes only in the sections parallel to the sample surface, while the sections normal
to the surface showed mainly black dots, indicating that the butadiene sequences were arranged in cylindrical formations aligned
parallel to the surface. Similar patterns were found by Matsuo
and co-workers in their BSB block copolymers. However, in these
samples it has not been established by bidirectional sectioning
that the observed stripes represent cylindrical formations.

Transitions between the various types of formations observed
in isolated cases have been found by Matsuo (65) in a series of
SBS block copolymers with varying composition. At high styrene
content, the butadiene sequences formed spherical particles. As
the styrene content decreased, cylindrical formations, and,
eventually, layered formations of the butadiene sequences were
found. Both types of formations appear as patterns of black

stripes in the photomicrographs; their exact geometry has to be
established by bidirectional sectioning. Samples at the opposite
end of the composition scale, i.e. at low styrene content, were
not studied. However, comparing this investigation with the
results obtained by Hendus et al. (62) and Inoue et al. (66) from
similar series of samples which showed formation of spherical
particles of the styrene sequences at low styrene contents, one
can speculate that the transitions from one formation to the other
as the composition is varied are probably analogous at both ends
of the composition scale. This would mean that the following
formations occur as the styrene content of the block copolymers
decreases and the butadiene content increases: butadiene spheres;
butadiene cylinders; alternate lamellae of butadiene and styrene;
styrene cylinders; styrene spheres. Recent thermodynamic calcula-
tions by Inoue et al. (80) support this sequence of geometries
over the entire composition scale.

Transition from one type of geometry to another within one
sample can also be made by changing the nature of the solvent,
from which the sample is prepared by evaporation. Morphological
changes occurring as a result of changes in solvent "goodness"
follow the lines established by the extensive studies on the behavior of BG copolymers in selective solvents reviewed above.
Some studies of solvent effects on styrene-butadiene block copolymers are included in several of the papers discussed in this
section (33,40,64,67,68).

After comparing the published morphological studies, one
comes to the conclusion that a correlation between the various sequential arrangements in block copolymers and the morphology of
the samples has not been established clearly, as yet. Approaching
the problem from the aspects of polymer compatibility and solution
behavior, one could speculate that the arrangements SB, BSB, and
SBS should have the same morphology, but some of the photomicrographs seem to indicate morphological differences between these
groups of block copolymers. Comparing the physical properties of
the three groups, it is obvious that profound differences exist
between samples of different block structure. Particularly the
studies of Matsuo and co-workers (65) indicate that, according to
physical properties, the block copolymers fall into two categories. The SBS and SBSB block copolymers, in the first category,
are tough materials. They resemble high-impact polystyrenes in
that they show stress-whitening. Even though they contain two
distinct phases, they are often transparent, because the phases
are so finely divided that the heterogeneities do not scatter
visible light (65). The SB and BSB block copolymers, in the
second category, are often as brittle as polystyrene (65). They
show much less stress-whitening, and while they are sometimes
transparent, they are more often translucent or opaque.

Styrene-butadiene block copolymers of the SBS type have found important commercial applications. They are thermoplastic elastomers which combine the properties of crosslinked, filled rubbers with the processability of conventional thermoplastic polymers. A series of thermoplastic rubbers has been developed by the Shell Chemical Company and is offered under the Shell trade names KRATON® and THERMOLASTIC® (6-8). The novel properties of these materials are apparently a direct consequence of the two-phase structure and of the special morphology of these block copolymers. A theoretical explanation of the observed physical properties has been offered by Holden (71a) and Meier (71b) in their concept of "domain" formation. According to this theory the incompatibility of the styrene and butadiene sequences leads to the formation of spherical domains of the styrene sequences, and the block copolymer molecules become arranged such that one styrene sequence of a molecule is part of one domain, while the other styrene sequence is part of another domain (71). The styrene domains thus act as physical crosslinks between the butadiene chains and behave simultaneously as hard filler particles. This gives the material the mechanical properties of a crosslinked, filled rubber. The view that the two styrene sequences of individual block copolymer molecules belong to separate domains also has been expressed by Hendus et al. (62). This structural model offers a very direct and plausible explanation of the physical behavior of SBS block copolymers and takes into account that apparently only SBS type block copolymers show this behavior. The only drawback of this model seems to be the postulate that each of the two styrene sequences of individual block copolymer molecules must belong to a different domain. Since there are no repulsive interactions between like polymer chains, there is no obvious reason that two styrene sequences should always segregate in this manner. Since the block copolymer molecules are flexible coils rather than stiff rods, one can easily visualize that many molecules fold back so that both styrene sequences enter the same domain.

In the recent literature, a number of papers on the preparation, characterization, and physical properties of SBS block copolymers have appeared (64-71). Some of these papers contain thorough discussions and analyses of the domain model. Particularly, the work of Beecher, Marker, Bradford, and Aggarwal (64) and Childers and Kraus (68) adds greatly to the understanding of this mode. Both groups find evidence for the existence of a long range rigid network in SBS block copolymer. The existence of this network is particularly evident in the stress-strain curves, which show a yield point in a first elongation cycle but not in a second elongation cycle, indicating that a network structure breaks down under stress in the first cycle. The network reforms on annealing at elevated temperatures or on storage of the samples at room temperature for a period of several months. The formation of the network seems to be a feature of the proposed cubic lattice, which

has been discussed above with reference to the same authors.
Studying a sample of KRATON 101, which had a styrene content of
28 weight percent and a number average molecular weight of 76,000,
Beecher and co-workers (64) came to the conclusion that the basic
morphological unit of KRATON 101 is a spherical polystyrene domain
with a diameter of 120 Å. However, the spherical domains are apparently not isolated from each other. As suggested by electron
microscopic evidence, they seem to be interconnected by tiny channels of polystyrene, thus giving rise to the formation of the
postulated long range, rigid network, i.e. the cubic lattice mentioned above.

A model of this nature not only serves as a basis for the
interpretation of the stress-strain curves, but it also eliminates
the need for the complicating requirement that the styrene sequences of individual block copolymer molecules occupy different
domains. By the same token, the model still leaves room for ques-
tioning why SBS block copolymers behave differently from BSB or
SB block copolymers. Considering the limited scope of presently
available data, this question may not be a real one, because the
possibility cannot be dismissed that the discrepancies between the
two categories of block copolymers are not so much an indication
of principal differences but are an illusion created by an attempt
to interpret data that are not complete. On one hand, it has not
been conclusively established that the presently known behavior of
SBS block copolymers is typical of all possible compositions and
morphologies. On the other hand, most of the SB samples investigated so far had layered structures, and the possibility cannot be
dismissed that block copolymers of this category might behave like
filled, crosslinked rubbers, if their composition were chosen such
that a spherical morphology were obtained. Discussing the problems
involved, one has to keep in mind that the understanding of block
copolymer morphology as a function of composition and structure is
in its beginning, and an understanding of the morphology of graft
copolymers in relation to structural parameters is nonexistent.
It appears that the papers reviewed in this article provide the
first and only available data on which attempts to arrive at such
correlations can be based. Therefore, the suspicion may not be
far-fetched that there are still undiscovered factors which give
rise to specific morphological differences between the various
categories of block copolymers. Knowledge of these factors will
be necessary for a complete interpretation of the physical behavior
of styrene-butadiene block copolymers and block and graft copolymers in general.

FORMATION OF PHYSICAL CROSSLINKS IN POLYURETHANES AND IONOMERS

The formation of physical crosslinks in the styrene domains
of styrene-butadiene block copolymers has interesting parallels

in other systems reported in the recent literature, namely in certain urethane polymers and in the so-called ionomers. Even though
these systems are not block copolymers in the ordinary sense,
domains of certain segments of the polymer chains seem to be formed
by principles analogous to those operating in styrene-butadiene
block copolymers: segregation of dissimilar segments and aggregation of similar segments.

Cooper, Tobolsky, and co-workers (72-75) published recently
a series of papers on the physical behavior of urethane polymers
such as SPANDEX® and ESTANE® urethane elastomers. Included in the
studies were also KRATON styrene-butadiene block copolymers. The
work started with the observation of unusual properties in the
SPANDEX system which were particularly evident in the modulus-
temperature curves. Two distinct glass transitions were found in
both the urethane elastomers and the KRATON styrene-butadiene
block copolymers studied for comparison. In the investigation of
styrene-butadiene block copolymers, the observation of distinct
styrene and butadiene transitions generally has been made and
always has been associated with the presence of two distinct phases
(57, 64-69). Accordingly, Cooper and co-workers came to the conclusion that their poly(ester-urethane) elastomers are segmented
linear polymers, i.e. that they are, in essence, block copolymers.
Like the styrene-butadiene block copolymers, the urethane block
copolymers behave like filled, crosslinked rubbers, and do so
apparently for the same reason. Cooper et al. explained the observed behavior on the basis of associations of rigid segments in
the polymer chains in domains which act as crosslinks and as
filler particles. The soft segments in the chains also associate
and influence the low-temperature properties of the elastomers
(72). The response under stress of the urethane block copolymers
indicates the breakdown of a network similar to the breakdown
reported by Beecher et al. (64) for styrene-butadiene block copolymers. This breakdown of a structure also occurs in ordinary,
carbon-black filled rubbers, in which the structure is apparently
the result of certain bonds between the carbon-black particles
(75).

The formation of another type of physical crosslinks has recently been reported by MacKnight and co-workers (76-78) and by
Longworth and Vaughan (79). These crosslinks are of an ionic
nature and occur in the so-called ionomers. Ionomers are metal
salts of copolymers of hydrocarbon monomers with monomers containing acid groups, e.g. salts of ethylene-methacrylic acid copolymers (79). The major distinction between ionomers and polyelec-
trolytes is that the ionic comonomer units in the ionomers are
only a minor portion of the entire polymer chain, while polyelec-
trolytes consist mainly of ionic monomer units. Thus, ionomers
are essentially hydrocarbons.

The aspect of the work on ionomers of interest in the context of the present article is the formation of ionic clusters in the ionomers. These clusters seem to play a similar role as the domains of styrene sequences in styrene-butadiene block copolymers and could be called "ionic domains." The formation of ionic domains by forming the salts from the acid form of the copolymers has a profound influence on the morphology and on the physical properties of the materials. The tensile strength and the melt flow viscosity of the copolymers increase markedly, and the materials become clear and transparent, while they were hazy and translucent before salt formation. Both research groups cited above are in agreement with respect to the importance of the domain formation in the explanation of the physical and morphological phenomena observed. They postulate a three-phase model and support this model by electron microscopic evidence and x-ray analysis. According to this model, ionomers consist of two polymer phases and an ionic phase. The amorphous and the crystalline polyethylene portions form the polymer phases and the clusters of ions form the ionic phase. In analogy to the results obtained with styrene-butadiene as well as urethane block copolymers, the ionic clusters seem to act like physical crosslinks and, maybe, filler particles in the polyethylene matrix. Like many of the styrene-butadiene block copolymers, the ionomers are clear and transparent, even though they contain more than one phase.

CONCLUSIONS

The research on colloidal and morphological behavior of block and graft copolymers has developed gradually from a few isolated observations to a sizeable body of knowledge. The first discovery was apparently Merrett's (1) observation that grafted natural rubber has the capability of forming micelles and organic sols and occurring in two modifications. After this original discovery, papers appeared sporadically. In recent years, the research activity in this area has increased significantly, mainly because new methods of synthesizing and characterizing BG copolymers have become available and accepted.

The most fruitful new synthetic approach has been the preparation of block copolymers by anionic polymerization. For the first time, block copolymers with relatively well-defined structure and composition became available in high yields and larger quantities. Used mainly with the commercially important monomers styrene, butadiene and isoprene, but also with the monomer pair styrene-ethylene oxide, which is of special interest because of the capability of the ethylene oxide chains to crystallize, anionic polymerization methods have started the recent wave of progress. Yet, most of the new results would not have been

obtained without the introduction of three methods of characterization into the field: phase microscopy, electron microscopy
in combination with the osmium tetroxide staining method, and
x-ray analysis, particularly low-angle x-ray scattering.

Surveying the data which recently have become available, it
is now possible to outline a scheme which describes the colloidal
behavior of block and graft copolymers in solution and in the
solid state. Since BG copolymers consist of homopolymeric substructures, it is mandatory to include in such a scheme considerations on the compatibility of polymers. Much speculation must be
included, because the amount of confirmed information is still
very limited. Particularly the treatment of graft copolymers together with block copolymers is largely speculation which assumes
that simple graft copolymers with only a few side chains behave
in essentially the same way as block copolymers. This assumption
is the more problematic, the more complex the structure of graft
copolymers becomes, and it must be kept in mind that almost nothing
is known about the influence of the fine structure of graft copolymers on their colloidal behavior.

The colloidal behavior of BG copolymers in solution is based
on two types of interactions: polymer-solvent interactions and
polymer-polymer interactions. In selective solvents, in which BG
copolymer molecules have a dual solubility, the colloidal behavior
is mainly determined by interactions between the polymer chains
and the medium. An example of this effect is the stabilization of
organic latices. In nonselective solvents, in which all portions
of the BG copolymer molecules are uniformly soluble, interactions
between different polymer chains become predominant and play the
main part in determining the colloidal behavior. This effect is
illustrated in polymeric oil-in-oil emulsions, in which immiscible
polymer solutions are emulsified by BG copolymers.

Both types of interactions can give rise to the formation of
oriented structures in solution and can determine the morphology
of solid samples formed by evaporation of the solvent, or by polymerization of the solvent if the solvent is also a monomer. In
selective solvents, three forms of orientation have been established: spherical micelles, cylinders, and lamellae. Cylindrical
and lamellar formations have first been found in solutions of
styrene-ethylene oxide copolymers in selective solvents, which are
apparently polymeric analogues of liquid crystals. Since ethylene
oxide chains are capable of crystallization, the observation of
mesomorphic structures in these systems has at first been attributed to the crystallizability of the ethylene oxide chains. However, more recent results obtained with block copolymers of styrene
with butadiene or isoprene, none of which can crystallize, indicate
that solutions of these block copolymers also in nonselective

solvents have properties found in liquid crystals: they exhibit
intense iridescence, usually with blue and green color tones.
One can thus speculate that polymer-polymer interactions, i.e.
the incompatibility of different polymer subchains, cause orientation of block copolymers in solution, and that the employment
of selective solvents or the use of polymer subchains which can
crystallize are not indispensable prerequisites for this type of
behavior, even though they may enhance it.

The types of morphological structures of BG copolymers in
solution in nonselective solvents have not been established conclusively. One can only speculate that the same types of structures exist, depending on various conditions, which have been
found in solutions prepared with selective solvents, namely
spherical micelles, cylinders, and lamellae. All of these structures now have been found in block copolymers in the solid state,
and it is not unreasonable to speculate that the morphology in
the solid state is an image of the morphology in solution in nonselective solvents, because it is known from solutions of different homopolymers in mutual solvents that the solvent plays mainly
the role of a diluent in these systems, but has no effect on the
phenomenon of polymer-polymer incompatibility per se.

A generalized scheme of the solid state morphology of BG copolymers is shown in Fig. 5. The existence of three geometries,
spheres, cylinders, and lamellae, has been confirmed for block
copolymers of styrene with either butadiene or isoprene. The main
variable determining which of the three types is formed seems to
be the volume ratio of the two phases. The dimensions, i.e. the
diameters of the spheres or cylinders or the distances between
adjacent lamellae, cylinders, or spheres, are determined by two

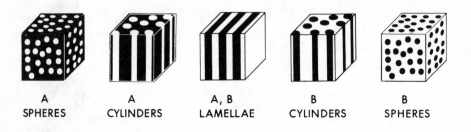

| A | A | A, B | B | B |
| SPHERES | CYLINDERS | LAMELLAE | CYLINDERS | SPHERES |

Increasing A-Content

Decreasing B-Content

Fig. 5. Scheme of block copolymer morphology. The cylindrical forms can be distinguished microscopically
from the lamellar form only by bidirectional
slicing of the specimens.

variables. In the absence of homopolymers, the lengths of the
subchains seem to be the main parameters determining the dimensions,
but in the presence of homopolymers, the total volume ratio of the
phases plays the dominant role. This suggests that also the composition of the block copolymers, which determines the phase volume
ratio in the absence of homopolymers, would have an influence on
the dimensions.

A scheme outlining the colloidal behavior of BG copolymers
would not be complete without some remarks about the compatibility
of polymers. The claim has often been made, particularly in the
patent literature, that BG copolymers have a "compatibilizing"
effect on blends of homopolymers. In the context of the present
knowledge on the morphology of BG copolymers, especially in the
solid state, this claim becomes essentially a question of semantics.
If "compatibility" is equated with "clarity to the naked eye" or
simply with "improvement of certain physical properties," then BG
copolymers can be said to compatibilize blends of homopolymers.
On a molecular level, however, there can be no compatibility between unequal polymer chains, and a "compatibilizing" effect, in
the strict sense of the word, cannot exist.

The question of compatibility of copolymers of any type with
the corresponding homopolymers is legitimate only in the case of
random copolymers. BG copolymers are neither compatible nor incompatible with the corresponding homopolymers. On a molecular
level, they are themselves mixtures of homopolymers, even though
the mobility of the subchains is restricted by chemical bonds.
When any of the parent homopolymers is added to a BG copolymer, it
blends into the corresponding phase, which already exists, thus
altering the morphology and the degree of dispersion, but not the
nature of this phase.

ACKNOWLEDGMENT

The stimulating discussions with Dr. T. Alfrey, Jr. on
various aspects of block and graft copolymers and particularly
his suggestions on the subject of intramolecular phase separation
are gratefully acknowledged. The critical proof-reading of the
manuscript by Dr. J. K. Rieke is much appreciated.

REFERENCES

1. F. M. Merrett, Trans. Faraday Soc., 50, 759 (1954).
2. F. M. Merrett, Ric. Sci., 25, 279 (1955).
3. F. M. Merrett, British Patent 797,346 (1958).
4. G. E. Molau, J. Polymer Sci., A, 3, 1267 (1965).

5. G. E. Molau, J. Polymer Sci., A, 3, 4235 (1965).
6. G. Holden and R. Milkovich, U. S. Patent 3,265,765 (1966).
7. J. T. Bailey, E. T. Bishop, W. R. Hendricks, G. Holden, and N. R. Legge, Rubber Age, 98, 69 (1966).
8. R. D. Deanin, SPE J., 23, 45 (1967).
9. W. J. Burlant and A. S. Hoffman, "Block and Graft Copolymers," (Reinhold, New York, 1960).
10. R. J. Ceresa, "Block and Graft Copolymers," (Butterworths, Washington, 1962).
11. H. A. J. Battaerd and G. W. Tregear, "Graft Copolymers," (Interscience, New York, 1967).
12. M. Szwarc, M. Levy, and R. Milkovich, J. Am. Chem. Soc., 78, 2656 (1965).
13. S. Schlick and M. Levy, J. Phys. Chem., 64, 883 (1960).
14. P. Rempp, V. I. Volkov, J. Parrod, and C. Sadron, Bull. Soc. Chim. France, 919 (1960).
15. G. Greber and J. Tölle, Makromol. Chem., 53, 208 (1962).
16. G. Greber, Makromol. Chem., 101, 104 (1967).
17. I. R. Schmolka and A. J. Raymond, J. Am. Oil Chem. Soc., 42, 1088 (1965).
18. A. Skoulios, G. Finaz, and J. Parrod, Comp. Rend., 251, 739 (1960).
19. A. Skoulios and G. Finaz, Comp. Rend., 252, 3467 (1961).
20. G. Finaz, A. Skoulios, and C. Sadron, Comp. Rend., 253, 265 (1961).
21. A. Skoulios and G. Finaz, J. Chim. Phys., 59, 473 (1962).
22. H. Bartl and W. v. Bonin, Makromol. Chem., 57, 74 (1962).
23. W. v. Bonin and W. Göbel, Kunststoffe, 53, 741 (1963).
24. J. Danon, M. Jobard, M. Lautout, M. Magat, M. Michel, M. Riou, and C. Wippler, J. Polymer Sci., 34, 517 (1959).
25. M. H. Jones, Canad. J. Chem., 34, 948 (1956).
26. F. D. Hartley, J. Polymer Sci., 34, 397 (1959).
27. I. E. Climie and E. F. T. White, J. Polymer Sci., 47, 149 (1960).
28. S. Ye. Bresler, L. M. Pyrkov, S. Ya. Frenkel, L. A. Laius, and S. I. Klenin, Vysokomol. Soed. 4, 250 (1962).
29. Y. Gallot, M. Leng, H. Benoit, and P. Rempp, J. Chim. Phys., 59, 1093 (1962).
30. F. M. Merrett, J. Polymer Sci., 24, 467 (1957).
31. G. E. Molau, to be published.
32. Y. Gallot, E. Franta, P. Rempp, and H. Benoit, J. Polymer Sci., C, 4, 473 (1963).
33. G. E. Molau, Macromolecules, 1, 260 (1968), presented at the Symposium on Polymeric Detergents, 154th ACS Meeting, Chicago, Sept. 10-15, 1967.
34. A. Dondos, P. Rempp, and H. Benoit, J. Chim. Phys., 62, 821, (1965).
35. A. Dondos, P. Rempp, and H. Benoit, J. Polymer Sci., B, 4, 293 (1966).

36. A. E. Skoulios, G. Tsouladze, and E. Franta, J. Polymer Sci., C, 4, 507 (1963).
37. C. Sadron, Chim. Pure et Appl., 4, 347 (1962).
38. C. Sadron, Angew. Chem., 75, 472 (1963).
39. C. Sadron, Chim. Ind., 96, 507 (1966).
40. B. Gallot, R. Mayer, C. Sadron, Comp. Rend., 263, 42 (1966).
41. A. Dobry and F. Boyer-Kawenoki, J. Polymer Sci., 2, 90 (1947).
42. R. L. Scott, J. Chem. Phys., 17, 279 (1949).
43. P. J. Flory, "Principles of Polymer Chemistry," Cornell University Press, Ithaca, N. Y., 1953.
44. W. H. Stockmayer, L. D. Moore, Jr., M. Fixman, and B. N. Epstein, J. Polymer Sci., 16, 517 (1955).
45. G. E. Molau, J. Polymer Sci., B, 3, 1007 (1965).
46. M. Lautout and M. Magat, Z. Phys. Chem., 16, 292 (1958).
47. G. M. Burnett, P. Meares, and C. Paton, Trans. Faraday Soc., 58, 737 (1962).
48. J. R. Urwin and J. M. Stearne, Makromol. Chem., 78, 194, 204 (1964).
49. S. Krause, J. Phys Chem., 68, 1948 (1964).
50. J. R. Urwin and J. M. Stearne, Eur. Polymer J., 1, 227 (1965).
51. H. Inagaki, Makromol. Chem. 86, 289 (1965).
52. H. Inagaki and T. Miyamoto, Makromol. Chem., 87, 166 (1965).
53. K. Kato, Polymer Eng. Sci., 7, 38 (1967).
54. G. E. Molau and H. Keskkula, J. Polymer Sci. A-1, 4, 1595 (1966).
55. G. E. Molau and H. Keskkula, Canad. Patent 754,636 (1967).
56. J. F. Henderson, K. H. Grundig, and E. Fischer, presented at the 13th Canadian High Polymer Forum, Ottawa, Sept. 22-24, (1965).
57. R. J. Angelo, R. M. Ikeda, and M. L. Wallach, Polymer 6, 141 (1965).
58. G. Riess, J. Kohler, C. Tournut, and A. Banderet, Rev. Gen. Caoutchouc., 3, 361 (1966).
59. G. Riess, J. Kohler, C. Tournut, and A. Banderet, Makromol. Chem., 101, 58 (1967).
60. J. Kohler, G. Riess, and A. Banderet, Eur. Polymer J., 4, 173, 187 (1968).
61. E. Vanzo, J. Polymer Sci., A-1, 4, 1727 (1966).
62. H. Hendus, K. H. Illers, and E. Ropte, Kolloid-Z., 216-217, 110 (1967).
63. E. B. Bradford and E. Vanzo, J. Polymer Sci., A-1, 6, 1661 (1968); J. Polymer Sci., C, 26, 161 (1969), presented at the Symposium on Block Copolymers, California Institute of Technology, Pasadena, June 5, 1967.
64. J. F. Beecher, L. Marker, R. D. Bradford, and S. L. Aggarwal, Polymer Preprints 8, 1532 (1967), presented at the Symposium on Polymer Modification of Rubber and Plastics, 154th ACS Meeting, Chicago, Sept. 13-15, 1967.

65. M. Matsuo, T. Ueno, H. Horino, S. Chujyo, and H. Asai, Polymer 9, 425 (1968); M. Matsuo, Japan Plastics, July 1968, pp. 6-16.
66. T. Inoue, T. Soen and H. Kawai, Polymer Letters, 6, 75 (1968).
67. R. A. Livigni, L. Marker, G. Shkapenko, and S. L. Aggarwal, presented at the Symposium on Structure and Properties of Elastomers, Meeting of the ACS Division of Rubber Chemistry and the Chemical Institute of Canada, Montreal, May 2, 1967.
68. C. W. Childers and G. Kraus, Rubber Chem. Technol., 40, 1183 (1967).
69. N. Canter, Polymer Preprints 8, 61 (1967).
70. R. E. Cunningham and M. R. Treiber, J. Appl. Polymer Sci., 12, 23 (1968).
71. a) G. Holden, E. T. Bishop, and N. R. Legge, J. Polymer Sci., C, 26, 37 (1969).
    b) D. J. Meier, J. Polymer Sci., C, 26, 81 (1969).
    presented at the Symposium on Block Copolymers, California Institute of Technology, Pasadena, California, June 5, 1967.
72. S. L. Cooper and A. Tobolsky, Textile Res. J., 36, 800 (1966).
73. S. L. Cooper and A. Tobolsky, J. Appl. Polymer Sci., 10, 1837 (1966).
74. S. L. Cooper and A. Tobolsky, J. Appl. Polymer Sci., 11, 1361 (1967).
75. S. L. Cooper, D. S. Huh, and W. J. Morris, Ind. Eng. Chem. Res. and Dev., 7, 248 (1968).
76. W. J. MacKnight, L. W. McKenna, and B. E. Read, J. Appl. Phys., 38, 4208 (1967).
77. W. J. MacKnight, T. Kajiyama, and L. McKenna, Pol. Eng. Sci., 8, 267 (1968).
78. W. J. MacKnight, L. W. McKenna, B. E. Read, and R. S. Stein, J. Phys. Chem., 72, 1122 (1968).
79. R. Longworth and D. J. Vaughan, Nature, 218, 85 (1968).
80. T. Inoue, T. Soen and H. Kawai, Preprints of papers presented at IUPAC Symposium on Macromolecular Chemistry, Toronto, 1968.
81. E. H. Andrews, Proc. Royal Soc., A 277, 562 (1963).

DILUTE SOLUTION PROPERTIES OF

BLOCK POLYMERS

R. SIMHA & L. A. UTRACKI

DIVISION OF MACROMOLECULAR
SCIENCE, CASE WESTERN RESERVE UNIVERSITY,
CLEVELAND, OHIO 44106 AND GULF OIL CANADA
RESEARCH CENTRE, STE-ANNE-DE-BELLEVUE, QUEBEC

INTRODUCTION

Dilute solution properties of high polymers can be adequately described by a two parameter[1] expression containing short range and long range interaction parameters A and B respectively. In the absence of long range interactions $B \equiv 0$ and the solution is at the θ-temperature. At this temperature the size of the polymer coils is solely defined by A.

In order to extend the two parameter theory to a copolymer system, one has to define the θ-condition. There are two possibilities [2,3]: a) it is possible to apply the same definition as for homopolymer solutions and specify that at $T=\theta$ the second virial coefficient $A_2=0$, i.e. that the overall long range interaction parameter $B_c=0$, (where the subscript refers to the copolymer) or b) one may demand that, under the θ-condition, all individual long range interaction parameters B_{ij} vanish identically. If the first definition is used, the θ-temperature for a block polymer will be a function of the size and number of individual blocks, the composition and total molecular weight M_c of the copolymer, in addition to the chemical nature of the solvent and solute species. As the polymer-polymer interactions usually are very strong, the condition $B_c=0$ actually results from a compensation of large interaction forces hence, large perturbances of the copolymer coils must be expected.

On the other hand, the second definition although experimentally unattainable, defines the unperturbed dimension of a copolymer coil as a unique quantity, analogous to that of a homopolymer. Our method of evaluation of dilute solution data of copolymers is based on the second definition.

METHOD OF ANALYSIS[2,3]

Consider a block copolymer composed of a series of chemical species $i=1,2,\ldots,n$. For species i let the molecular weight be M_i, the molecular weight per carbon atom in the chain backbone, M_{oi}, the unperturbed mean square end-to-end distance, $<r_\theta^2>_i$, the short range interaction parameter, $A_i^2 = a_i^2/M_{oi} = <r_\theta^2>_i/M_i$, and the long range interaction parameter $B_i = \beta_i/M_{oi}^2$, where β_i is the so-called binary cluster integral of polymer i in the solvent. If x_i is the mole fraction of the species i, we can write for a copolymer:

$$M_{oc} = \sum_i M_{oi}$$

$$A_c^2 = a_c^2/M_{oc} \qquad\qquad (1)$$

$$a_c^2 = \sum_{ij} (a_{ii}^2 X_{ii} + a_{ij}^2 X_{ij})$$

$$a_i = A_i^2 M_{oi} = <r_\theta^2>_i (M_{oi}/M_i)$$

where X_{ij} is the appropriate probability of an i–j contact along the chain backbone, and the summation, extends over all chemical species. With a proper evaluation of X_{ij}, the relation should be valid for any copolymer-solvent system.

For block polymers we have:

$$X_{ij} << X_{ii}$$

$$a_c^2/M_{oc} = \sum_i a_i^2 x_i/M_{oc}$$

The long range interaction parameter B_c for a copolymer solution can be written as:

$$B_c = \beta_c/M_{oc}^2 \tag{2}$$

$$\beta_c = \sum_{ij} \beta_{ij} \xi_{ij}$$

where ξ_{ij} is the fraction of i-j configuration characterized by a parameter β_{ij}. An explicit compution of ξ_{ij} in a highly perturbed state is not easily performed. We shall instead explore the validity of the simplest possible assumption, namely a random distribution of disconnected segments in the volume occupied by the perturbed copolymer coil. In this case we have:

$$\xi_{ij} = x_i x_j \tag{2a}$$

$$\beta_c = \sum_i \beta_{ii} x_i^2 + 2 \sum_{i<j} \beta_{ij} x_i x_j$$

If this assumption is valid, the β_{ij} should turn out to be independent of x_i and M_i and vary only with the solvent for a given polymer. In this manner equations (1), (1a) and (2a) extend a two parameter relation developed for homopolymers in solution to a copolymer-solvent system.

The randomness of segment distribution in the perturbed state is the most precarious assumption of our method. For a system in which this randomness does not occur, one must expect variations of the polymer-polymer long range interaction parameters with composition and molecular weight. Most probably this will be the case for a copolymer dissolved either in a solvent which is a precipitant for one block, or in a mixture of chemically different solvents.

Equations (1), (1a) and (2a) will now be applied to the following polymer-polymer-solvent systems:

I-Styrene-Butadiene, SB and SBS block polymers[2]

II-Styrene-Methyl Methacrylate,SM and MSM block polymers[3]

III-Styrene-Methyl Methacrylate,statistical copolymers[3]

IV-Styrene-Methyl Acrylate,statistical copolymers[4]

As for system I, the polymerization of butadiene (B) and styrene (S)-butadiene di (SB) and tri (SBS) blocks has been carried out in benzene at $30^{\circ}C$, using n-butyllithium as the initiator[2]. The characteristics of the samples are shown in Table I.

Table I

Characteristics of Polymer Samples

Sample	$M_k \times 10^{-3}$	$M_n \times 10^{-3}$	$W_k \%PS$	$W\%PS$	x
PS-1	78.1[a]	78	–	–	–
PS-2	147[a]	137	–	–	–
PS-3	222[a]	201	–	–	–
PB-1	ca. 20	17.8	–	–	–
PB-2	18	28.6	–	–	–
PB-3[b]	95	33.0	–	–	–
PB-4	66	74.4	–	–	–
PB-5	39	75.6	–	–	–
PB-6[c]	ca.120	156	–	–	–
PB-7	160	423	–	–	–
SB-1	14.3 + 28.6	43.1	33.3	32.9	0.1131
SB-2	14.3 + 92.7	105	13.4	17.8	0.0533
SB-3	14.3 + 133	129	9.71	12.3	0.0352
SB-4	14.3 + 571	620	2.44	3.59	0.0096
SBS-1	7.5 + 18 + 7	34.8	44.6	45.9	0.1811
SBS-2	10 + 24 + 10	44.8	45.5	45.5	0.1783
SBS-3	7.5 + 28.6 + 8.8	54.8	36.3	36.6	0.1299
SBS-4	10 + 49 + 10	69.4	29.0	30.1	0.1005
SBS-5	10 + 100 + 10	117	16.7	20.7	0.0634
SBS-6[b]	14 + 62 + 14	141	31.1	40.2	0.1486
SBS-7	13 + 150 + 13	170	14.8	11.6	0.0329
SBS-8	10 + 450 + 9	517	4.05	4.05	0.0108

a M_n values reported by McCormick

b Sample extracted from a partially insoluble batch of branched material.

c Crack in the reactor prior to completion of the polymerization

An effort has been made to keep the size of the S-block constant, thus varying the composition by a change in size of the B-segments. Dilute solution viscosity and osmotic pressure measurements at 34.2°C in toluene, dioxane and cyclohexane were performed. The results are plotted according to the following equations for the intrinsic viscosity and the second virial coefficient:

$$[\eta]/M_n^{\frac{1}{2}} = \phi(A_\eta^3 + 0.51 B_\eta M_n^{\frac{1}{2}}) \tag{3}$$

$$A_2 M_n^{\frac{1}{2}} = 1.65\times10^{23} A_\Pi^3 + 0.968\times10^{23} B_\Pi M_n^{\frac{1}{2}} \tag{4a}$$

$$A_2 M_n^{\frac{1}{2}} = 2.83\times10^{23} A_\Pi^3 + 1.67\times10^{23} B_\Pi M_n^{\frac{1}{2}} \tag{4b}$$

as is seen in Figs. 1 and 2

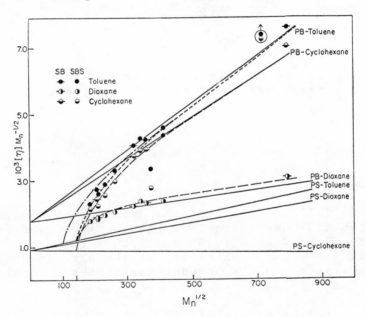

Fig.1. Molecular weight dependence of intrinsic viscosity: (——) experimental data for homopolymers and (● , ◐ , ◖) for copolymers; (---) from eqs.(1a), (2a) and (3) for SBS in three solvents; (---) for SB in toluene

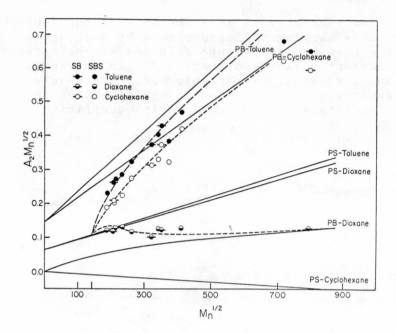

Fig.2. Molecular weight dependence of A_2: (——) data
for homopolymers; points for copolymers.

The full lines represent the data for homopolymers.
As the chemical composition for the copolymer varies
(from 2 to 33 wt % of S in SB and from 4 to 45 wt % of
S in SBS), the relative location of the experimental
points in respect to the homopolymer curves naturally
also changes

The numerical values of the $\mathring{a}$ and β parameters
for styrene and butadiene used in the subsequent cal-
culations are given in Table II. They were determined
by us and agree very well with the values quoted in
the literature. The smoothed values of $\overline{\beta}_{SB}$ quoted at
the bottom of the Table are indicated by the solid
lines in Fig.3.

Table II

Interaction Parameters

Parameter	Equation	Toluene		Dioxane		Cyclohexane	
		PS	PB	PS	PB	PS	PB
$K_\theta \times 10^3$	(3)	0.920	1.76	0.920	1.76	0.920	1.76
$a, \overset{o}{A}$	(3)	4.9	3.1	4.9	3.1	4.9	3.1
	(4a)	5.3	3.5	5.3	–	–	3.5
	(4b)	4.5	3.0	4.5	–	–	3
$\beta, \overset{o}{A}{}^3$	(3)	3.9	0.89	3.0	0.20	-0.20	0.78
	(4a)	8.4	1.6	8.3	–	–	1.33
	(4b)	4.9	0.94	4.8	–	–	0.77
$\bar{\beta}_{SB}, \overset{o}{A}{}^3$	(3)	3.2		1.9		2.7	
	(4a)	5.3		–		–	
	(4b)	3.1		–		–	

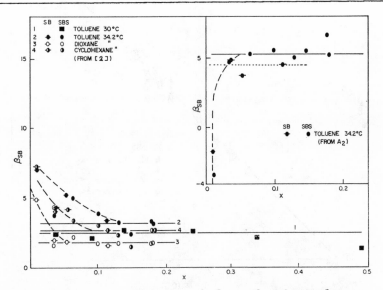

Fig.3. Parameter β_{SB} calculated from $[\eta]$ in toluene,
cyclohexane, and dioxane, and from A_2 in toluene
at 34.2°C. Values calculated for Shell SBS
samples from $[\eta]$ in toluene at 30°C are also shown.

Note that no differences between SB and SBS data can be detected. For x<0.1 an increase of β_{SB} is observed. This increase is more pronounced for solutions in toluene and cyclohexane, both good solvents for B-blocks, than in dioxane. In the last solvent only one point, for the lowest x-value deviates. Now in our systems, the molecular weight M_c of the copolymer increases with decreasing x. In the range $x<0.1, M_c>10^5$, a progressive increase of polydispersity and branching as well as a lowering of accuracy of the osmotic pressure data should be expected. In addition, it has been previously reported[2] that in this range of M_n, the osmotic pressures recorded by a high speed membrane osmometer (such as the one used in the study) are too high.

In the same figure we also plot β_{SB} computed from $[\eta]-M_c$ data for a series of Shell SBS samples with con-

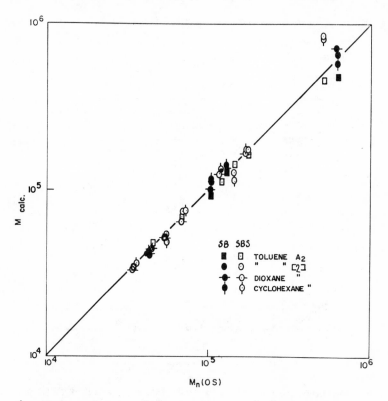

Fig.4. Molecular weights computed from eqs. (1), (2a), (3) and (4b) vs. M_n determined from osmotic pressure.

stant M_c and the composition varying between 13 and 80 wt % of styrene. Note that for this series of data, the polymer-polymer interaction parameter is independent of x over the whole range of compositions.

Using the smoothed values of $\bar{\beta}_{SB}$, we can now re-compute the molecular weights of the polymers. The results are shown in Fig.4. The quantities M_{calc} are de-rived from intrinsic viscosities and should correspond to viscosity averages. For samples SB-4 and SBS-8, the ratio M_{calc}/M_n=1.13 and 1.63, which is quite reasonable a value for M_n<500,000. Interestingly enough, these polymers also exhibit the largest variations of β_{SB} with composition. A similar computation for the Shell samples yields $M_{calc}=M_n\pm7\%$.

Turning now to System II, copolymers of styrene and methyl methacrylate, SM and MSM, have been most frequently investigated with widely differing results. The intrinsic viscosities in benzene and toluene for a number of SM and MSM samples as determined by various authors are summarized in Table III[3]. The values of β_{SM} calculated from these data are shown in Fig.5 as a function of the molar styrene content x.

We note that the points computed from the data of the same laboratory do not show any systematic varia-tion with x. This systematic change should be expected if the block polymer in solution exists in the form of separated styrene and methyl methacrylate domains as earlier postulated by other authors. In fact, the data points cluster into three groups. For each of these, the values of β_{SM} can be related to the method of poly-merization of the M-block and, as such, to the tacticity of the M-segment. The heterotacticity index for the middle group is 39-48%, whereas for both bottom and top groups it is ca.5%. However, the method of polymeri-zation of the samples of the lowest group was reported to yield gelled solutions, characteristic of the high content of long syndiotactic diads of M-blocks. This did not occur in the polymerization of the samples for which β_{SM}-values cluster around the upper-most line.

In Fig.5 there are included the β_{SM} values compu-ted for four statistical SM copolymers, (Group III considered here). These are also found to be independ-ent of x, and to lie within the group of values computed for block polymers with large heterotacticity of M-segments.

Table III

Intrinsic Viscosities and Number Average
Molecular Weights of Di-(SM) And Tri-(MSM)
Block Copolymers

No.	M_n $\times 10^{-3}$	x	Solvent and temp[b]	$[\eta] / M_n^{\frac{1}{2}} \times 10^3$	$\beta_{12} \overset{o}{A}^3$
1	229	0.631	1	1.73	3.77
2	259	0.550	1	2.83	11.38
3	316	0.490	1	2.54	8.33
4	278[a]	0.712	II	1.10	-2.33
5	529[a]	0.783	II	1.44	-1.64
6	1150[a]	0.591	II	1.56	-0.32
7	2170[a]	0.243	II	0.95	-2.65
8	187	0.470	I	2.05	7.57
9	225	0.611	I	2.18	7.45
10	302	0.692	I	2.32	7.27
11	390	0.845	I	2.40	8.30
12[c]	400	0.641	III	2.11	4.29
13[c]	603	0.510	III	1.97	2.47
14[c]	345	0.500	III	2.03	4.47
15	36	0.450	I	1.09	2.20
16	37.4	0.302	I	1.05	2.31
17	36	0.450	I	1.09	2.20
18	317	0.490	I	1.62	2.51
19	509	0.420	I	2.05	3.89
20	530	0.470	I	2.11	3.89

[a]M_{app} from light scattering. [b]I, toluene,
25.0°; II toluene, 30°; III, benzene, 30°.
[c]Diblocks.

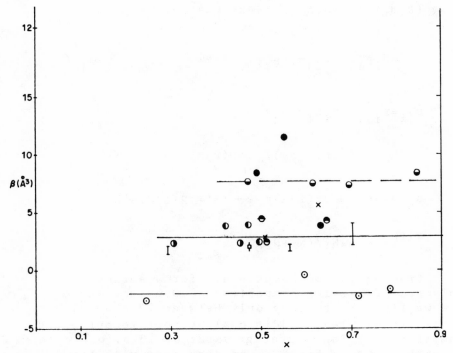

Fig.5   PS-PMMA interaction parameter β_{12} as a function
        of mole fraction of PS.   Points and crosses
        represent values for block copolymers computed
        directly (See Table III) and interpolated from
        Figure 7, respectively.   The bars indicate the
        range of β_{12} computed for statistical copolymers
        in three solvents[3].   The square represents β_{12}
        for butanone solution with a standard deviation
        indicated by the bar.

        Whereas the computations of β_{12} for block and
statistical copolymers can be made by means of the same
equation, the unperturbed dimensions of the two polymers
depend on composition in a different way.   In the case
of statistical copolymers, the number of polymer 1 -
polymer 2 contacts along the chain backbone, see eq.(1),
must be taken into account.   Their contribution to a_c
can be calculated from the statistics of copolymerization, in terms of the composition and (for a low
conversion system) simply the reactivity ratios.   The
pertinent relations are summarized below[3].

$$(N_1 + N_2) a_c^2 = a_1^2 N_{11} + a_2^2 N_{22} + 2a_{12}^2 N_{12}$$

$$N_{ij} = N_i p_{ij}, \quad p_{ij} = 1 - p_{ii}$$

$$p_{ii} = r_i n_i / (r_i n_i + n_j)$$

$$N_1 / N_2 = n_1 (r_1 n_1 + n_2) / \quad n_2 (r_2 n_2 + n_1)$$

$$a_c^2 = a_1^2 x p_{11} + a_2^2 (1-x) p_{22} + 2x(1-p_{11}) a_{12}^2$$

with $x = N_1 / (N_1 + N_2)$

(5)

From eq.(5) we compute a_{12} for a series of copolymers of varying composition. Over the whole range we find a_{SM}^2 to vary only between 23 and 23.9 $\mathring{A}^2$. This result lends additional support to our basic assumptions. Using an average value of 23.5, one obtains the curve shown in Fig.6. The agreement with the experimental data is certainly satisfactory, considering experimental errors.

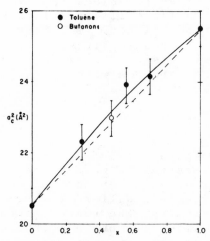

Fig.6. Parameter a_c^2 of PS-PMMA statistical copolymer in toluene (filled circles) and
butanone (open circle), 30°, as a function
of the mole fraction of PS. The solid
line was computed from eq.5. Bars indicate the standard deviation computed from
the original [η]-data.

Table IV   STYRENE-METHYL ACRYLATE INTERACTION PARAMETERS

Samples	St.mole fraction, x	a_{12}^2 a_c^2, $Å^2$ exp.	calc.	Benzene β_1	β_{12} $Å^3$	Toluene β_1	β_{12} $Å^3$	Butanone β_1	β_{12} $Å^3$
PMA	0.000	—	17.3	2.00	—	0.683(?)	—	1.42	—
SM-25	0.258	18.5	18.3	—	8.63	—	—	—	4.00
SM-40	0.425	19.3	19.4	—	5.95	—	3.92	—	2.44
SM-50	0.527	19.9	19.6	—	—	—	—	—	—
SM-65	0.658	19.3	20.3	—	6.44	—	6.34	—	3.78
SM-76	0.776	21.0	21.1	—	7.73	—	—	—	3.84
PS	1.000	22.0	—	2.84	—	2.84	—	0.46	—
Averages:		19.3±1.1	—	—	7.19	—	—	—	3.51

β_{12} (benzene,butanone) ±15% $Å^3$

It is worth noting that the average value of a_{12}^2 is very close to the arithmetic mean for the parent polymers, viz. 20.5 and 25.5, or 23.0$\overset{\circ}{A}^2$.

Finally, the same analysis can be performed for the dilute solution data of the styrene-methyl acrylate statistical copolymer[4]. (Group IV considered here). The results are shown in Table IV.

Also in this case we observe no systematic variation of a_{12}^2 with x. The average value of this parameter, 19.3$\pm$1.1, again is close to the mean value for homopolymers (19.6). Using the experimental average value of a_{12}^2 we may compare computed and experimental values of a_c^2. The agreement is excellent, the largest difference being 5% for polymer SM-65.

The values of β_{12} derived for benzene and butanone solutions, also do not vary with composition in any systematic way; the standard deviation of the average values was 15%. The data for the toluene solutions are too meager for a meaningful analysis.

CONCLUSIONS

Judging from the constancy of the a_{12}'s and β_{12}'s, the proposed method of evaluation of solution data of block and statistical copolymers describes adequately all four series analyzed. This suggests that the polymer segment distributions are random, and a significant separation of homopolymeric blocks into domains does not occur. However, one must bear in mind that in all systems investigated the solvents were either good for both polymeric components, or good for one and approximately θ-solvents for the other.

The equations used to describe a two component block polymer-solvent system contain the following parameters: a_1, a_2, β_1, β_2, characteristic of the respective homopolymers, and β_{12}, x and M_c. As the last two parameters can be derived from the kinetics of polymerization, a determination of the polymer-polymer mutual interaction parameter solely is required for the complete description of the system.

It is noteworthy that for all systems studied by us, β_{12} is a positive and numerically large, comparable to β_i for a homopolymer in a good solvent. This is contrary to what one would expect from the phase equilibria of polymer-polymer-solvent systems. There incom-

patibility or phase separation is the general phenomenon. The incompatibility at higher concentrations
and the large positive values of β_{12} may indicate a
strong concentration dependence of the entropic component of this long range interaction parameter.

We show finally in Fig.7 a correlation of β_{12}
with β_2 for the SM statistical copolymer in five
solvents[3].

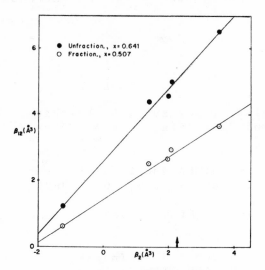

Fig.7   PS-PMMA interaction parameter β_{12} vs.
        PMMA-solvent interaction parameter β_2.
        Open and filled circles represent values
        for fraction and original unfraction-
        ated polymer, respectively. The arrow
        indicates β_2 for PMMA-benzene at 25°.

A similar pattern has also been observed for SB and
SBS block polymers[2], and for S-methyl acrylate statistical copolymers[4], both in three different solvents. It is interesting to note in contrast, that
the variations of β_S do not appear to correlate with
β_{12}.

The values of a_{12}^2 computed for S-methyl methacrylate and S-methyl acrylate statistical copolymers were found to be very close to the mean value
of a_1^2 and a_2^2. Recently, Wunderlich[5] reported the

same observation for the methyl methacrylate-n-butyl-acrylate statistical copolymer. Such results are indicative of a special relationship between the hindrance potentials of the two polymeric compounds and will depend on the nature of the pendant groups. In fact, from Shimura's data[6] for a styrene-acrylonitrile statistical copolymer we obtain $a_{12}^2 = 25\text{Å}^2$, whereas the mean equals $(a_1^2 + a_2^2)/2 = 23.1$. Similarily, for the α-methyl styrene-ethylene pair $a_{12}^2 = 27\text{Å}^2$, whereas $(a_1^2 + a_2^2)/2 = 20\text{Å}^2$.

REFERENCES

1. See for example: T. Norisuye, K. Kawahara, A. Teramoto and H. Fujita, J. Chem. Phys., **49**, 4330 (1968)

2. L. A. Utracki, R. Simha and L. J. Fetters, J. Polymer Sci., Part A-2, **6**, 205 (1968)

3. L. A. Utracki and R. Simha, Macromolecules, **1**, 505 (1968)

4. H. Matsuda, K. Mamano and H. Inagaki, J. Polymer Sci., Part A-2, **7**, 609, (1969)

5. W. Wunderlich, Macromolecular Colloquium, Freiburg/Br.(Germany), Febr. 27 - March 1, 1969

6. Y. Shimura, J. Polymer Sci., Part A-2, **4**, 423 (1966)

7. H. Tanazawa, T. Tanaka and A. Soda, J. Polymer Sci., Part A-2, **7**, 929 (1969)

MECHANICAL PROPERTIES OF STYRENE-ISOPRENE BLOCK COPOLYMERS

R.A. Robinson[*+] and E.F.T. White[*]

[*]Department of Polymer and Fibre Science

The University of Manchester Institute of Science

and Technology, Manchester, England

[+]BP Research Centre, BP Chemicals Limited

Sunbury-on-Thames, Middlesex, England

INTRODUCTION

Certain segmented block copolymers prepared by anionic polymerisation techniques have recently attracted considerable attention because of their unusual mechanical behaviour in exhibiting
long range rubber elasticity without the necessity for chemical
crosslinking[1-6]. Commercial materials in which a block of rubber
is embedded between two blocks of a glassy material, such as polystyrene, to give a two-phase "domain" structure have been widely
investigated,[7,8] but few systematic studies of the general relationships between structure and properties have been undertaken for this
class of polymers. The sequence size and arrangement, the overall
molecular weight and molecular weight distribution, the microstructure and the morphology will all have considerable effects on
these relationships and there is a strong need for these to be
clarified.

At present it appears that the high tensile strength of commercial segmented block copolymers of the styrene-butadiene-styrene
type arises from the presence of discrete micro-domains of polystyrene surrounded by a rubbery matrix,[4,9] which act as deformable
energy-absorbing filler particles and also serve as junction points
for rubber chains. Both the associated glassy segments and the

flexible components appear to retain their separate chain motions and these are associated with separate glass-rubber transition temperatures for the two components[10]. It is, however, not yet apparent to what extent these glass temperatures are affected by the incorporation of segments into a block copolymer structure on its subsequent separation into a two-phase domain morphology.

In this paper we have attempted to prepare a comprehensive range of styrene-isoprene-styrene (S-I-S) and isoprene-styrene-isoprene (I-S-I) block copolymers of well characterised structure and to make comparative measurements of their transition temperatures and mechanical behaviour and relate these to their molecular and phase structure.

EXPERIMENTAL

Segmented block copolymers of both the S-I-S and I-S-I types having varied compositions were prepared by anionic polymerisation employing the "living" polymer technique. The disodium tetramer of α-methylstyrene was used as initiator[11] and polymerisations were carried out in tetrahydrofuran solution at -40°C by sequential addition of the appropriate monomers to the rapidly stirred initiator solution. Initiator and monomer concentrations were calculated to give copolymers of similar overall molecular weight. The resulting block copolymers were precipitated into methanol containing 2.0% phenyl β-napthylamine as an antioxidant and dried in vacuum.

Molecular weights and molecular weight distributions were determined by gel-permeation chromatography and found to agree with the expected kinetic values. Copolymer compositions were similarly in agreement with the experimental results using ultra-violet absorptiometry. The microstructure of the polyisoprene segments was investigated by infra-red spectroscopy on solutions of the polymer in carbon disulphide using the procedure of Binder and Ranshaw[12]. Characterisation data for all the polymers are given in Table 1.

Glass-transition temperatures were determined from thermograms obtained with a Perkin-Elmer Differential Scanning Calorimeter (Model DSC 1) using a heating rate of 16°C/min. and at maximum sensitivity.

The mechanical response of the materials to free torsional oscillations around 1 Hz over a temperature range of -30°C to +110°C was measured using a Nonius torsion pendulum. Test specimens were prepared by compression moulding thin sheets (0.05 - 0.07 in. thick) at 135 - 170°C. Strips 0.25 in. wide and approximately 3 in. long were then cut from the sheets for torsional tests.

Table 1

Characterisation data for Styrene-Isoprene Block Copolymers

| Polymer No. | Wt. % Styrene | Molecular Weight x 10^{-3} (by G.P.C.) | | Molecular Weight Distribution | Microstructure of Polyisoprene Blocks |
		$\overline{M}_n$	$\overline{M}_w$		
ISI/98	97.4	103.5	119.0	1.15	
ISI/95	94.2	115.0	143.7	1.25	59.3% 3,4
ISI/90	88.0	113.0	148.1	1.31	
ISI/80	80.7	114.5	151.0	1.30	30.6% 1,2
ISI/60	64.2	106.0	172.0	1.62	
ISI/40	40.7	112.0	181.7	1.62	10.1% trans-1,4
ISI/20	20.7	100.6	128.0	1.27	
SIS/4	95.9	.106.7	127.0	1.19	
SIS/10	90.7	102.3	121.0	1.18	59.3% 3,4
SIS/15	85.1	98.5	114.2	1.16	
SIS/20	79.0	104.0	118.9	1.14	30.6% 1,2
SIS/40	57.9	108.5	127.0	1.17	
SIS/60	41.5	102.0	117.5	1.15	10.1% trans-1,4
SIS/80	18.6	95.5	119.4	1.25	

Uniaxial stress-strain measurements on parallel sided samples approximately 1 mm. thick were obtained at 20°C using an Instron at various strain rates.

Fracture surface energies (γ) were measured on parallel-sided samples containing induced natural edge-cracks of various lengths (c). Samples were extended to rupture at a strain rate of 0.1 min^{-1}. at 20°C and the true ultimate stress (σ_{ult}) calculated from the load-elongation curve and the cross-sectional area of the sample at the fracture or tearing plane. The Griffith criterion[13]

$$\sigma_{ult} = \left(\frac{2E\gamma}{\pi c}\right)^{\frac{1}{2}}$$

in which E is the tensile modulus was used to calculate the fracture surface energy γ. This quantity is considerably larger than the theoretical surface energy since it includes a large energy contribution from irreversible deformations in and near the surface.

Morphological details of the micro-phase structure were
obtained by transmission electron microscopy, sufficient contrast
being provided by osmium tetroxide staining of the polyisoprene
phase. Ultra thin (500 - 1000Å) sections were cut from epoxy-
embedded samples.

RESULTS AND DISCUSSION

(a)   Transition Behaviour

The styrene-isoprene block copolymers investigated are characterised by two discrete transition regions indicative of each polymer
segment type. An upper glass-transition temperature of about 100°C
corresponds to the transition of the polystyrene segments while a
lower transition at around 10°C is assumed to indicate the onset of
rotation of the polyisoprene segments. Examination of the homopolymers show similar transition temperatures. This clearly suggests
a two phase system, a conclusion consistent with the known immiscibility of polystyrene and polyisoprene. The relatively high transition temperature observed for the polyisoprene blocks is due to the
high concentration of 3,4- and 1,2 structures in sodium initiated
polyisoprene prepared in a polar solvent. These branched structures
would be expected to raise the main chain transition temperature from
that for the cis-1,4 structure. Increasing the degree of external
unsaturation in polyisoprenes increases the Tg, and while it is not
yet possible to prepare the pure 3,4 isomer, extrapolation of available glass-transition data on materials of up to 90% external
unsaturation would indicate a transition temperature of some 50°C
for the 3,4 structure. This is consistent with the value of 50°C
for 3 methyl but-1-ene, a similar type of structure.

The absolute values of the glass-transition temperatures are,
however, for terblock polymers of similar overall molecular weight
dependent on copolymer composition. Figure 1 shows the glass-
transition temperature of both polystyrene and polyisoprene blocks
as a function of composition. The introduction of any polyisoprene
segment to form a block copolymer causes an immediate decrease in
the styrene transition temperature. This reduction is larger in
S-I-S blocks as here the polyisoprene block not only introduces a
rubbery region into the polymer matrix, but also halves the polystyrene chain length. If there was a complete phase separation
with no interaction at the interface then the initial reduction of
the polystyrene Tg would not be expected. This behaviour may be
explained by assuming some partial compatability of polystyrene and
polyisoprene regions in an environment where the glassy polystyrene domains are connected to rubbery polyisoprene chains by
primary valence bonds. Further increases in polyisoprene block length

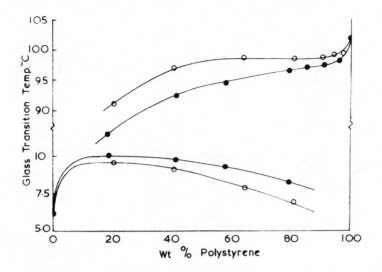

Figure 1.   Glass-transition temperatures of styrene/isoprene block copolymers.   ——o——   I-S-I blocks   ——●——   S-I-S blocks.

at constant overall molecular weight, results in a continuing reduction in the styrene transition temperature consistent with the shorter segment length of the polystyrene blocks.

The transition temperatures of the polyisoprene blocks in the copolymers exhibit a similar but reverse effect. Here the presence of stiffer polystyrene blocks raises the transition of the poly-isoprene segments by an amount which decreases as the polyisoprene block length decreases. It seems reasonable, if some compatability between the differing polymer segments is presupposed, to expect the transition region of the rubber blocks to be shifted to higher temperatures when chemically bound to rigid glassy domains. Similar behaviour has been found for graft copolymers consisting of poly-styrene segments grafted on a polyisoprene backbone[14].

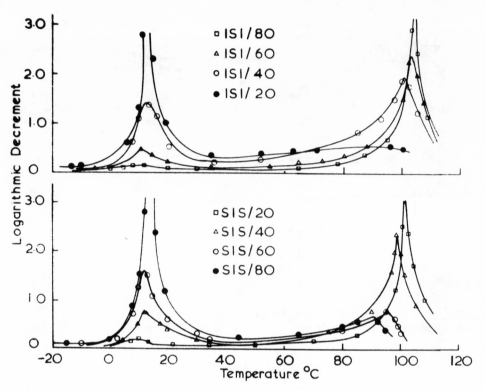

<u>Figure 2</u>   Temperature and composition dependence of logarithmic decrement for styrene/isoprene block copolymers.

The variation of mechanical damping with temperature for both block copolymer sequences, Figure 2, shows a similar transition dependence on copolymer composition. Damping maxima were obtained in proportion to the quantity of each block type present and the width of the damping peaks show a relatively broad transition region when compared to the homopolymers. There is some evidence that a polystyrene block with isoprene terminal blocks finds it more difficult to achieve a suitable phase separated structure than the corresponding S-I-S structure, particularly at low styrene concentrations. These effects can also be attributed to some partial compatability between polystyrene and polyisoprene segments. Canter[15] has found some similar results for S-B-S copolymers. He found that the width of absorption peaks could be considerably reduced by orienting the material, presumably assisting in the phase separation process.

(b)   Mechanical Properties

Mechanical properties of block copolymers are particularly susceptible to the manner of specimen preparation and to the previous mechanical and thermal history of the specimen. In addition to the usual difficulties in preparing a molded isotropic sheet there will be further uncertainties concerned with the overall nature of the phase morphology. This may change for example from a discontinuous to a continuous lamella structure depending on the particular preparative conditions employed, with corresponding large changes in mechanical behaviour. Similar difficulties can occur with solvent cast materials. Also the morphological structure may vary from place to place within a sample due to the heterogeneous character of its thermal history. Accordingly in any comparison of mechanical behaviour, the samples should be carefully prepared and the morphology controlled.

Measurements of the shear moduli, obtained using the Nonius torsion pendulum, of block copolymers having both central and terminal styrene segments are as shown in Figure 3.

Comparison of the shear moduli of block copolymers of similar composition but of differing structural arrangement reveals some variance in mechanical behaviour.

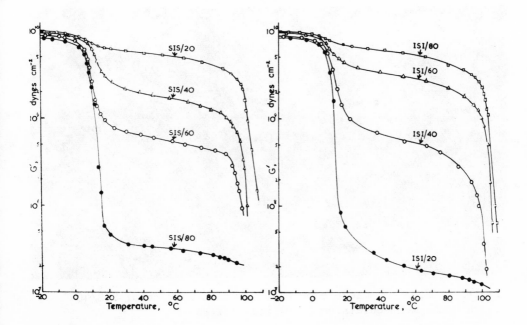

Figure 3 Variation of shear modulus (G') of block copolymers of styrene and isoprene with composition and temperature.

For the materials of high styrene content there is little
difference in the modulus-temperature spectrum between the two
samples. In each case electron photomicrographs show that the polystyrene phase is continuous with spherical polyisoprene inclusions.

At lower styrene contents, the ISI/60 shows greater rigidity
over the whole temperature range than SIS/40. The former material
has a continuous polystyrene matrix while the S-I-S copolymer has
a lamella structure with continuous regions of each phase as shown
in Figure 4.

At 40% styrene contents both types of block copolymers are
lamella like in phase structure, but the S-I-S arrangement is stiffer,
expecially at higher temperatures on account of the better crosslinking and entanglement features in the polyisoprene phase effected
by the styrene end blocks.

Similarly block copolymers containing about 20% polystyrene
indicate the stiffer characteristics of the S-I-S arrangement at
temperatures between the transition temperatures of the component
blocks. Both types of material, ISI/20 and SIS/80 exhibit a domain
type morphology with spherical polystyrene aggregates embedded in a
rubbery matrix giving rise to physical cross-linking and modulus
reinforcement in the case of the S-I-S material.

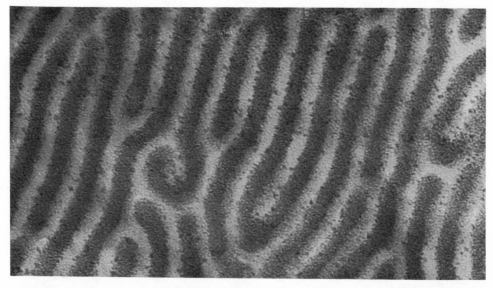

Figure 4   Electron micrograph of ultra thin section of SIS/40
block copolymer containing 57.9% polystyrene.

0.1 μ

The initial tensile moduli of styrene-isoprene block copolymers are, at ambient temperature, considerably dependent on the speed of testing. This is illustrated in Figure 5 which shows the values of Young's modulus (E) at different strain rates. All copolymers exhibit greater moduli at higher strain rates though polymers of high poly-isoprene content show considerably more reinforcement and at the highest strain rates are plastic in behaviour. The latter result is due to the close proximity of the test temperature (20°C) to the glass transition of the polyisoprene segments. As an increase in the speed of testing is equivalent to decreasing the test temperature, a higher tensile modulus of the polyisoprene phase would be expected at high strain rates.

In comparing the curves for I-S-I and S-I-S blocks it is noticeable that, when the polyisoprene segments exist as a con-tinuous phase, copolymers containing terminal polystyrene segments are stiffer especially at low strain rates. This can again be explained by considering the structure of the domain morphology in such block copolymers. In blocks containing terminal polystyrene segments, the latter form domains which then act both as network junctures for crosslinking and as a reinforcing filler. In I-S-I blocks the polystyrene domains are not chemically linked by poly-isoprene chains and no crosslinking of this nature can occur.

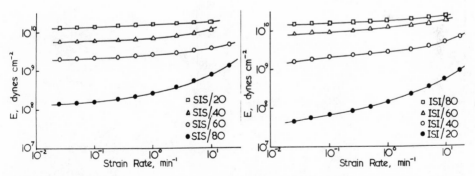

Figure 5   Variation of tensile modulus (E) with strain rate for various styrene/isoprene block copolymers at 20°C.

Stress-strain curves for block copolymers of both sequence
types obtained at various strain rates are plotted in Figures 6
and 7. As expected the tensile strength increases and ultimate
elongation decreases with increasing styrene content. As the styrene
fraction is reduced the failure behaviour of the various copolymers
changes from a brittle fracture to a yield point followed by drawing
to an elastic extension. At high rubber levels (above 80% polyisoprene the samples have low moduli and can be extended to high
largely reversible elongations. After extension the rubbery copolymers exhibit some permanent set, the magnitude of which varies with
styrene content, sequence arrangement, extension ratio and other
factors. In copolymers with terminal styrene blocks the set is
small and much lower than in the corresponding I-S-I blocks. An
S-I-S copolymer containing 18.6% polystyrene, for instance, exhibited less than 5% permanent set after having been stretched to
400% strain at 0.5 min^{-1}. The comparable copolymer of inverse
block arrangement (ISI/20) showed a permanent set of 25% when

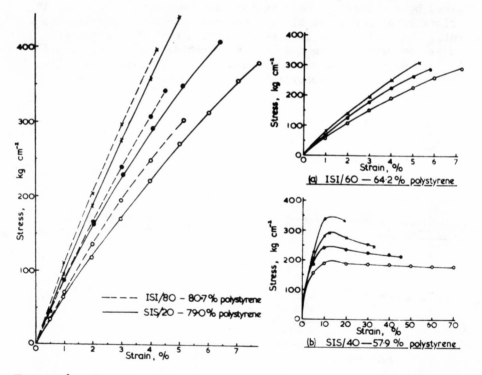

Figure 6. Tensile stress-strain curves for various block copolymers
of styrene and isoprene obtained at different strain rates at 20°C

—O— 0.05 min^{-1}.   —●— 0.5 min^{-1}.

—✕— 5.0 min^{-1}.   —△— 50 min^{-1}.

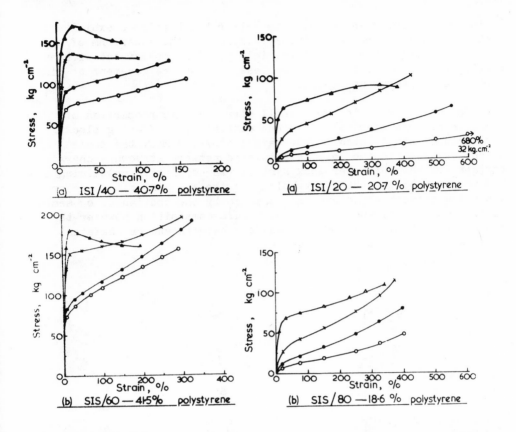

Figure 7. Tensile stress-strain curves for various block copolymers of styrene and isoprene obtained at different strain rates at 20°C.

—○— 0.05 min⁻¹.   —●— 0.5 min⁻¹.

—✕— 5.0 min⁻¹.   —△— 50 min⁻¹.

stretched under the same conditions. This clearly highlights the better network properties inherent in the S-I-S block system at this composition. On the other hand, at a styrene content of 40%, both sequence types show considerable permanent set. This can be explained by the presence of extensive regions of continuous aggregated styrene blocks which undergo irreversible plastic deformation when extended.

The stress-strain curves for materials containing about 60% styrene (ISI/60 and SIS/40) differ considerably, the material with terminal isoprene segments being brittle with low extensibility, while SIS/40 is of lower modulus and exhibits a characteristic yield point. This is consistent with the previously noted phase morphology.

It is interesting to note that the latter block polymer exhibits a
yield stress which bears a linear relation to the logarithm of the
strain rate suggesting that this yield point is associated with a
particular viscous deformation process such as the movement or disruption of polystyrene domains.

    The differences in the overall stress-strain properties of
block copolymers of similar styrene content but differing block
arrangements are related to two main factors, namely the phase
morphology and the two level elastomeric network structure characteristic of the S-I-S terblock system. In copolymers containing above
40% polystyrene the phase morphology is the most important factor
since the network structure is frozen in by the continuous or semi-
continuous polystyrene phase. Below this composition however the
rubber phase becomes continuous and the network and entanglement
effects become the major influence.

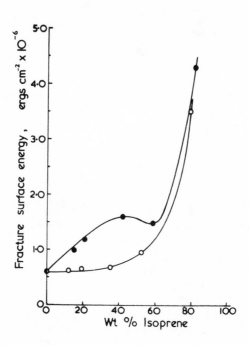

Figure 8. Dependence of the fracture surface energies of styrene/
isoprene block copolymers on copolymer composition at 20°C.

——O——   I-S-I blocks;   ——●——   S-I-S blocks.

The variation of fracture surface energies (γ) with composition is illustrated in Figure 8 for both S-I-S and I-S-I block arrangements. The fracture surface energy of homopolystyrene (6.1×10^5 ergs/cm^2) measured here is in agreement with values determined for this polymer by other methods[16]. As the styrene content decreases the fracture surface energy increases but is a surprising dependence on block arrangement, the S-I-S materials being substantially tougher when the polystyrene phase is continuous (from 40 - 100% styrene).

Bucknall and Smith[17] have shown that gross yielding of rubber modified plastics involves the formation of large numbers of crazes in the glassy phase, each craze initiating at the surface of a rubber particle and usually ending at the surface of another one. If a craze mechanism is responsible for the fracture of styrene-isoprene block copolymers containing a continuous polystyrene phase (and crazes have been observed in these materials) an increase in isoprene content would lead to the formation of shorter, more stable crazes because the inter-domain distance is smaller. Such copolymers would have a higher fracture energy.

Electron micrographs of the phase structure of S-I-S copolymers show that they possess a more ordered domain arrangement than the corresponding I-S-I material. At a specific composition then the crazes formed in an S-I-S block would be shorter and more stable than those in an I-S-I copolymer, and would lead to a higher fracture surface energy.

This explanation is consistent with the largest difference in occurring at approximately 60% styrene content, SIS/40 has a lamella like structure (Figure 4) and the craze length is limited to the distance between polyisoprene lamellae. (ca 175Å). In the ISI/60 material however morphology consists of a continuous polystyrene phase interdispersed with polyisoprene domains, and crazes of significantly greater length may be formed.

ACKNOWLEDGEMENTS

The authors wish to thank The British Petroleum Co. Ltd., for permission to publish these results and for the award of the Dunstan Research Fellowship to R.A.R. We are also indebted to Mr. S.T.E. Aldhouse for the molecular weight data and Mr. P.H. Drake for the electron micrographs.

REFERENCES

1.   A.W. van Breen and M. Vlig: Rubber and Plastics Age,
                                   47, 1070, (1966).

2.   J.T. Bailey, E.T. Bishop, W.R. Hendricks, G. Holden and
                                   H.R. Legge:
                                   Rubber Age, 98, 69, (1966)

3.   Shell Research: British Patent, 1,000,090   (1965)

4.   C.W. Childers and G. Kraus: Rubber Chem. Technol., 40, 1183,
                                   (1967).

5.   M. Matsuo, T. Ueno, H. Horino, S. Chujyo and H. Asai:
                                   Polymer, 9, 425, (1968).

6.   R.E. Cunningham and M.R. Treiber: J. Appl. Polymer Sci.,
                                   12, 23, (1968).

7.   J.F. Beecher, L. Marker, R.D. Bradford and S.L. Aggarwal:
                                   Polymer Preprints, 8, 1532, (1967).

8.   N.H. Canter: J. Polymer Sci., A-2(6), 155, (1968).

9.   M. Norton, J.E. McGrath and P.C. Juliano: J. Polymer Sci. Part C.,
                                   In press.

10. G. Kraus, C.W. Childers and J.T. Gruver: J. Appl. Polymer Sci.,
                                   11, 1581, (1967)

11. M. Szwarc, J. Smid and C.L. Lee: J. Phys. Chem., 66, 904, (1962).

12. J.L. Binder and A. Ranshaw: Anal. Chem., 29, 503, (1957).

13. A.A. Griffith: Phil. Trans. Roy. Soc., London, 221A, 163, (1920).

14. B.F. Ekin: M.Sc. Thesis, University of Manchester, 1968.

15. N.H. Canter: Polymer Sci., A2, 6, 155, (1968).

16. L.J. Broutman and F.J. McGarry: J. Appl. Polymer Sci.,
                                   9, 589, (1965).

17. C.B. Bucknall and R.R. Smith: Polymer, 65, 437, (1965).

TIME-DEPENDENT MECHANICAL PROPERTIES OF ELASTOMERIC BLOCK POLYMERS IN LARGE TENSILE DEFORMATIONS

Thor L. Smith

I.B.M. Research Laboratory

San Jose, California 95114

ABSTRACT

Consideration is given to the stress-strain and ultimate tensile properties of an unplasticized and a plasticized styrene-butadiene-styrene triblock polymer at temperatures from -40 to 60°C. It is concluded that the time dependence of their mechanical properties and also their high tensile strengths result from energy dissipation associated with the plastic deformation, and eventual disruption, of the colloidal polystyrene domains. A similar mechanism is undoubtedly responsible for the high strength of segmented diisocyanate-linked (polyurethane) elastomers and also styrene-butadiene rubber vulcanizates filled with colloidal polystyrene spheres. It is further concluded that a non-crystallizable elastomer will exhibit high strength provided it contains about 25% of colloidal plastic particles, uniformly dispersed. For optimum reinforcement, the particles should deform plastically to dissipate stored elastic energy, thus relieving unfavorable stress concentrations in the vicinity of slowly growing, or incipient, cracks.

INTRODUCTION

All polymeric materials that exhibit high strength and toughness contain a dispersed phase. For example, the toughest plastics and fibers are semicrystalline; rubber is dispersed in polystyrene to impart toughness; and reinforcing fillers are ordinarily incorporated in non-crystallizable elastomers because, without filler, they are weak except under special test conditions (1).

137

Recently, considerable interest has developed (2) in thermoplastic elastomers, exemplified by styrene-butadiene-styrene (SBS)
triblock polymers. When the styrene content is about 25% and the
molecular weight of the polymer is not unduly low, the material is
an exceptionally strong and tough elastomer, consisting of a rubbery
polybutadiene matrix in which aggregates (domains) of polystyrene
chains are dispersed. By anchoring the ends of the triblock molecules, the colloidal domains impart a permanent three-dimensional
structure, hence preventing viscous flow below the softening temperature of the domains. In addition, the domains function as non-
rigid filler particles, and chain entanglements in the polybutadiene
phase assume the role of network junction points.

Like the triblock polymers, segmented diisocyanate-linked
(polyurethane) elastomers, which are alternating block polymers,
exhibit high strength. Such polymers are ordinarily not crosslinked by primary valence bonds (3) but contain plastic domains
formed from the so-called hard segments (4). Other elastomeric
materials whose toughness results from dispersed domains are the
silicone-polycarbonate block copolymers (5-7) and the ionomers (8,
9), the latter being copolymers of α-olefins and carboxylic acids
in combination with mono- or divalent cations.

In this paper, the manner in which plastic domains affect the
mechanical properties and, in particular, impart high toughness and
strength are considered. Toward this end, a discussion is first
given of key results from a recent study (10) of two triblock
elastomeric materials, namely, Kraton 101 and Thermolastic 226
supplied by the Shell Chemical Company. Then, the ultimate properties of these materials are considered along with results from a
recent study (11,12) of styrene-butadiene rubber (SBR) vulcanizates
reinforced with colloidal polystyrene spheres. Finally, tensile
strength data on diisocyanate-linked elastomers are discussed.

MECHANICAL CHARACTERISTICS OF TRIBLOCK POLYMERS

Materials and Experimental

As discussed previously (10), Kraton 101 is a styrene-butadiene-styrene (SBS) triblock polymer whose $\overline{M}_n$ is about 76,000 and
which contains about 27% by volume of polystyrene. Thermolastic
226 is an SBS triblock polymer that contains about 11% of an in-
organic pigment and about 35% plasticizer. On a total volume
basis, the volume fractions of polystyrene and plasticizer are
about 0.20 and 0.38, respectively. Provided the plasticizer
remains entirely in the rubbery matrix (an assumption which undoubtedly is not strictly valid), the matrix contains nearly equal
amounts of polybutadiene and plasticizer.

Stress-strain and ultimate property data were determined with an Instron tester at seven temperatures from -40° to 60°C and at nine crosshead speeds from 0.02 to 20 inches per minute. The original stress-strain data were converted (10) into plots of log $\lambda\sigma273/T$ vs log t, where the points for each plot correspond to a fixed value of the extension ratio λ. The quantity $\lambda\sigma$ is the stress based on the cross-sectional area of a deformed specimen; T is the test temperature in °K; and the time t equals $(\lambda-1)/\dot{\lambda}$, where $\dot{\lambda}$ is the extension rate. Typical results on Kraton 101 are shown in Figure 1. The left and right panels show, respectively, data at $\lambda = 1.2$ and $\lambda = 5$. Data on Thermolastic 226 gave similar plots, as illustrated elsewhere (10).

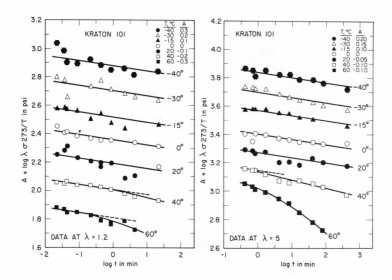

Figure 1   Stress-time data obtained at $\lambda = 1.2$ (left panel) and $\lambda = 5$ (right panel) from stress-strain curves measured at different extension rates and temperatures on Kraton 101

Relaxation Rate and Temperature Dependence

For both Kraton 101 and the plasticized Thermolastic 226, the data at $1.2 \leq \lambda \leq 5$ and at all temperatures except 40 and 60°C could be represented by parallel straight lines whose slope $(\partial \log \lambda\sigma/ \partial \log t)_{\lambda,T}$ equals -0.033; this value corresponds to a relaxation rate of about 8% per decade of time. For $\lambda > 5$, the relaxation rate increased significantly with extension. At 40 and 60°C, the relaxation rate for Kraton 101, and especially for Thermolastic 226, increased with time, especially at $\lambda > 5$, undoubtedly reflecting plastic flow and breakup of the domains which are relatively soft at these temperatures. Childers and Kraus (13) measured the stress relaxation at λ = 1.5 of an SBS triblock polymer of somewhat higher styrene content than Kraton 101. Between -60° and 25°C, their data show a constant relaxation rate of about 15% per decade of time.

Plots like those in Figure 1 were shifted along the log t axis to obtain values of log a_T, shown in Figure 2. Stress-time data at values of λ from 1.2 to 5 gave sensibly the same values of a_T. Significantly, the a_T data are the same for both Kraton 101 and Thermolastic 226.

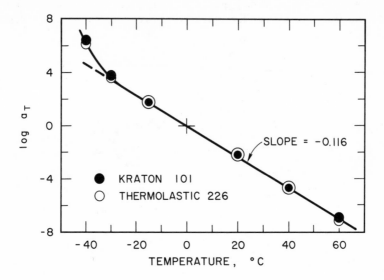

Figure 2   Values of log a_T for Kraton 101 and Thermolastic 226.

Ultimate Tensile Properties

For Kraton 101 below 40°C and for Thermolastic 226 below 0°C, the rupture stress and the elongation at break depend only slightly on temperature and extension rate. Because the domains are relatively soft at the other test temperatures, the ultimate properties depend quite strongly on extension rate and temperature.

Ultimate property data on the triblock materials at an extension rate of 1 min^{-1} are plotted against temperature in Figure 3 which also shows data on an unfilled styrene-butadiene vulcanizate (SBR-1) and an unfilled butyl rubber vulcanizate. Because the latter crystallizes readily under stress only below about 25°C, its strength and extensibility decrease markedly in a narrow temperature range above 25°C.

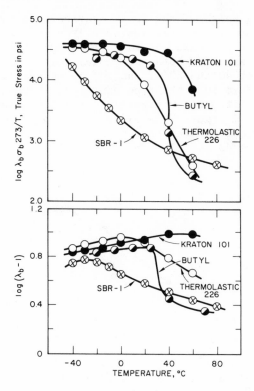

Figure 3   True tensile strength $(\lambda_b \sigma_b 273/T)$ and ultimate strain $(\lambda_b - 1)$ as a function of temperature.

DISCUSSION OF RESULTS

Dependence of Stress-Strain Data on Time and Temperature

The time and temperature dependence of the stress at a fixed value of λ can conceivably result from: (a) the slippage of entanglements in the rubbery matrix; (b) a nonaffine migration of the polystyrene domains; (c) the progressive plastic deformation of the domains; and (d) a breakdown of domain structure. Childers and Kraus (13) concluded that the polystyrene domains participate in the stress-relaxation process, either by nonaffine motions or by the detachment of polystyrene chains from the domains. From an examination of electron micrographs of stretched films, Beecher et al (14) proposed that, under a progressively increasing stress, the contorted cylindrical domains first break up into ellipsoidal units which then may break down by either ductile rupture or the continuous extraction of chains from the domains.

In considering which of the above-mentioned mechanisms has a
controlling effect on the relaxation rate, the key experimental
findings are: (a) the relaxation rate, $[\partial \log \lambda\sigma/\partial \log t]_{\lambda,T}$, at
$1.2 \leq \lambda \leq 5$ is the same for both the unplasticized and plasticized
materials, except at 40 and 60°C; and (b) the shift factor $\log a_T$,
which does not conform to a WLF-type equation, is the same for both
materials. The lowest plot in Figure 4, prepared using average
values of $\log a_T$ from Figure 2, shows that the data conform to an
Arrhenius equation, within the experimental accuracy, and give an
activation energy of 45.5 kcal.

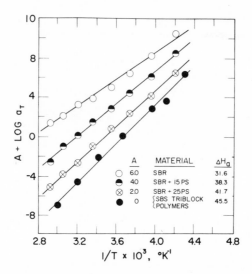

A	MATERIAL	ΔH_a
○ 6.0	SBR	31.6
◑ 4.0	SBR + 15 PS	38.3
⊗ 2.0	SBR + 25 PS	41.7
● 0	{SBS TRIBLOCK POLYMERS	45.5

Figure 4   Log a_T data for SBS triblock polymers (lowest line) and
            for SBR vulcanizates containing 0, 15 and 25% by volume of
            polystyrene spheres (data for SBR vulcanizates from Ref. 12).

For linear and lightly crosslinked polybutadiene polymers, and
also for two highly branched materials, $\log a_T$ data have been represented (15) by a WLF-type equation. Activation energies derived
from the equation are 7.0 and 11.4 kcal at 60 and -40°C, respectively.
These low activation energies, in contrast to 45.5 kcal for the
triblock polymers, indicate that the relaxation rate for the triblock
polymers is not controlled by slippage of entanglements in the polybutadiene phase. Furthermore, if entanglement slippage were the
dominant mechanism, the relaxation rate and the a_T data for Kraton
101 would undoubtedly differ from those for Thermolastic 226 whose
matrix contains plasticizer and polybutadiene in nearly equal amounts.

The addition of carbon black to an SBR vulcanizate has been
found (16) to increase the activation energy for relaxation processes from about 26 kcal for the unfilled vulcanizate to 40 kcal

for the one containing 30 parts of black. This increase was attributed to the nonaffine motion of carbon black particles, under an unbalanced force. The particles, being large compared with the surrounding polymer chains, encounter considerable viscous resistance in moving through the matrix. In the triblock polymers, the domains can, and probably do, move nonaffinely under certain conditions. However, such a mechanism undoubtedly does not control the relaxation rate because a nonaffine migration would occur more readily in the highly plasticized Thermolastic 226, hence effecting a reduction in activation energy. As such is not observed, it is concluded that the deformation and, under certain conditions, the disruption of domains are responsible for the time and temperature dependence of the mechanical properties.

Although bulk polystyrene, unless preoriented, does not deform ductilely on a macroscale, the plastic flow of polystyrene domains is quite reasonable; several contributing factors can be mentioned. Owing to the colloidal nature of the domains, their properties undoubtedly differ from those of bulk polystyrene. Also, the domains are under a thermal stress (predominantly compressive) because the thermal expansion coefficient of polybutadiene is about threefold greater than that for polystyrene. In addition, when a tensile specimen is stretched uniaxially, the domains experience a tensile stress in one direction and a compressive stress in the two perpendicular directions. Finally, it should be emphasized that conventional polystyrene can exhibit ductility on a microscale. By applying a tensile load, crazes have been produced in compression molded polystyrene and also in annealed sheets (17). As crazing involves cavitation and a subsequent plastic deformation of the surrounding material, it is apparent that polystyrene exhibits ductility under appropriate conditions.

The conclusion that the time and temperature dependence of mechanical property data, at least at large deformations, results from a plastic deformation of the polystyrene domains is consistent with the finding (18) that the activation energy for the viscous flow of an SBS triblock polymer approaches that for polystyrene below about 100°C. In the discussion elsewhere (10), the temperature dependence of a_T for the triblock polymeric materials is compared further with that for bulk polystyrene. Also, the important role played by the polystyrene phase is indicated by torsional pendulum data (19) obtained on a series of SBS polymers. For triblock polymers containing from 24 to 50% polystyrene, the reported data show that d log G'/dT = -0.0025 °C^{-1}, where G' is the storage modulus at 1 Hz, over the temperature range from -50 to 40°C. That is, the temperature dependence is the same regardless of whether the polybutadiene or the polystyrene form the continuous phase.

Except for the above-mentioned torsional pendulum data, the discussion thus far has been limited to large-deformation behavior.

Thus it should be pointed out that relaxation processes at small
deformations may differ from those at large deformations. Under
very small deformations, the stress on the polystyrene domains may
not be sufficiently large to cause yielding and plastic flow, except
at high temperatures. Such different behavior should affect the
a_T data. For a styrene-isoprene-styrene triblock polymer, data (14)
obtained with a vibrating reed apparatus gave a_T values that appear
to conform approximately to a WLF-type equation at temperatures
below about 0°C. More recent data (20), obtained on SBS polymers
in small deformations over extended ranges of temperature, frequency
and time, gave a_T values that conform quite closely to a WLF-type
equation below some temperature that lies in the range 30-60°C;
above this temperature, the data follow an Arrhenius equation. The
exact temperature below which the WLF-type equation is obeyed
depends on factors that as yet have not been resolved.

Origin of High Strength

There is considerable evidence (1,21) that rupture involves
three steps: (a) the formation of a crack or cavity, or the
"activation" of a pre-existent crack or cavity; (b) the slow growth
of a crack, commonly at a progressively increasing rate, until an
instability develops; and (c) the high-speed propagation of the
crack across the specimen in a direction perpendicular to the
maximum principal stress. The instability criterion for high-speed
crack growth can be considered to have the form $WC \geq K\mathcal{S}$, where W
is the elastic stored energy, C is the crack length, K is a dimen-
sionless geometrical factor, and $\mathcal{S}$, which has dimensions of
$ergs/cm^2$, can be termed the fracture surface energy. The magnitude
of $\mathcal{S}$ depends on the dissipative processes, and thus the detailed
mechanism, accompanying crack growth; these depend on material
characteristics and the speed of crack growth. For present
purposes, however, we need only consider that high-speed crack
growth, and hence macroscopic rupture, will occur when WC exceeds
some critical value, even though the critical value may depend
on experimental conditions.

In non-crystallizable elastomers that contain plastic domains,
the density of stored elastic energy may be reduced by the viscous
dissipation of energy in the rubbery matrix, by a plastic deformation and disruption of domains, and by cavitation in the vicinity
of the primary crack, i.e., the particular crack that eventually
propagates catastrophically.

For conventional vulcanizates, the flaw or crack at the instant
of instability is often considered (22,23) to have an effective
size in the range 0.01-0.1 mm. For sake of discussion, suppose
that the critical crack size in Kraton 101 is considerably smaller,
e.g., between 10^{-4} and 10^{-3} mm; even this size, however, is manyfold

larger than that of the domains. Thus, if a crack begins to develop
in the rubbery matrix, for example, its initial size will be at least
one or two orders of magnitude smaller than the assumed critical
value. When the crack grows until its tip approaches or reaches a
domain, the driving force for further growth will then be reduced,
or prevented from increasing further, by a plastic deformation of
the domain. Also, the local density of elastic stored energy (and
thus the stress concentration) may possibly be reduced by the formation of cavities under the triaxial tensile stress which may
become particularly severe when the stress field (triaxial) in
front of the crack overlaps the triaxial field in the vicinity of
domains. To satisfy the criterion WC $\geq$ K$\mathcal{B}$, not only must W be
large but also C must be manyfold greater than domain dimensions.
A crack will achieve a critical size only after it has disrupted
a large number of domains; to do so, considerably energy must be
dissipated in plastic flow and possibly in the formation of secondary cavities and cracks, which tends to impart stability. The
viewpoint is that the domains, being small in size and rather
closely spaced, impede crack growth. To obtain a crack of critical
size in a finite period, the elastic stored energy, and thus the
applied stress., must be large.

Tensile Strength of Block Polymers and Related Materials

     In the triblock polymers, the plastic domains are coupled to
the rubbery matrix through primary valence linkages. In the SBR
vulcanizates filled with polystyrene spheres, studied by Morton
et al, (11,12) only a few chemical linkages apparently existed
between the matrix and each polystyrene particle. One of the
prepared vulcanizates contained 25% by volume of spheres whose
diameters were about 350 A. Tensile strength data obtained (11)
on this material at an extension rate of 12.7 min^{-1} are shown in
Figure 5 to be essentially the same as those obtained on Kraton 101
at 1 min^{-1}. (In this figure, the tensile strengths are based on
the initial cross-sectional area of a specimen.) Although the two
materials were tested at different extension rates, this difference
is of minor importance because, except at the higher temperatures,
the tensile strengths of both materials are only slightly dependent
on extension rate (10,12). The fact that their tensile strengths
are equal supports the view that chemical attachments between
filler particles and a rubbery matrix do not directly affect the
tensile strength.

     The high tensile strengths of Kraton 101 and the SBR vulcan-
izates filled with polystyrene spheres undoubtedly result from a
similar mechanism for reinforcement. Consistent with this premise,
the activation energy from a_T data for the SBR vulcanizates increased
from 31.6 kcal for the unfilled material to 41.7 kcal for the
vulcanizate containing 25% by volume of polystyrene spheres (500 Å

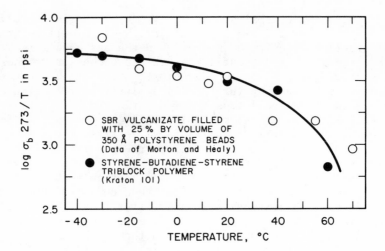

Figure 5   Tensile strength data of Kraton 101 compared with those
            of an SBR vulcanizate containing 25% by volume of poly-
            styrene spheres (latter data from Ref. 11).

in diameter), as shown in Figure 4.   (The a_T values, presented by
Morton et al (12), were obtained from the rate and temperature
dependence of the tensile strength.)

     There is, however, a significant difference between the prop-
erties of Kraton 101 and the polystyrene-filled SBR vulcanizates.
Following rupture, the "permanent" set for Kraton 101 is relatively
large compared to that for the SBR vulcanizates (24). This observation does not vitiate the contention that the high strength
results primarily from a plastic deformation of the polystyrene
spheres. (Admittedly, if the spheres were absolutely rigid, the
tensile strength quite probably would be increased; the magnitude
of the increase cannot be estimated, although it undoubtedly would
be considerably less than that observed.) The domains in Kraton 101
initially have a contorted cylindrical shape. Under a progressively
increasing stress, they deform and then break up into spherical
(more precisely, ellipsoidal) units which undergo a plastic deformation (14). When the stress is removed, a specimen can regain its
original dimensions only after the polystyrene units, under the
internal stress provided by the matrix, recover and form the initial
domain configuration. At ambient temperature, recovery proceeds at
a very slow rate.*

---

*A creep and creep-recovery test has been made (25) on Thermolastic
226. After three days under a constant load (unknown stress) at
ambient temperature, the extension was 94%; following removal of
the load, the specimen recovered to within 1% of its initial length
in seven days.

On the other hand, the SBR vulcanizates contain uniformly dispersed spheres whose diameter is greater than that of the contorted domains in Kraton 101. During a tensile test, it seems probably that few of the spheres are disrupted, although they undoubtedly deform plastically. Hence, the recovery immediately following a test should be greater than that for Kraton 101.

To conclude the discussion of tensile strength, the properties of segmented diisocyanate-linked elastomers will be considered briefly. A study has been made (26) of soluble (not chemically crosslinked) elastomers prepared from polyoxypropylene glycol (PPG), toluene-2,4-diamine (TDA) and toluene-2,4-diisocyanate (TDI). The upper formula in Figure 6 shows that the so-called hard, or polar, segment contains two substituted urea and two urethane moieties as well as three aromatic rings. Such a structure would hardly be soluble in the soft-segment material composed of polyoxypropylene glycol coupled by TDI. Apparently the hard segments segregate into domains, yielding a three dimensional network structure.

Figure 7 shows that the tensile strength, determined on various formulations at 49°C, is an increasing function of the weight percent of the plastic (hard) phase. In Figure 7, the column labeled NH_2/OH gives the relative molar proportions of TDA and PPG for a particular formulation; similarly, $NCO/(NH_2+OH)$ gives the molar ratio of TDI to the sum of the moles of TDA and PPG. The weight percent of the hard phase was calculated by assuming that it consists of the TDA and associated TDI moieties; the excess TDI, because it probably reacted with substituted urea moieties, was also considered to reside in the hard phase. One formulation was prepared from equal molar amounts of TDA, PPG, and TDI; its tensile strength at 25°C is shown by the triangle.

Figure 7 also shows data for some Adiprene L elastomers, which are prepared from polytetramethylene ether glycol, 4,4'-methylene-bis-(2-chloroaniline), designated MOCA, and TDI; the resulting structure is shown by the lower formula in Figure 6. Formulation and tensile strength data at 25°C are from the trade literature (27). For the formulations considered, the hard phase was estimated to range from 26 to 50% by weight; the hard phase was assumed to consist of the MOCA and the TDI moieties that terminate the diisocyanate liquid prepolymer (27). Although the reported values of σ_b and λ_b depend considerably on composition, the true tensile strength (Figure 7) for most compositions is between 2.4 and 3.2×10^4 psi, essentially the same tensile strength exhibited by the triblock polymers (Figure 3).

Because the Adiprene L elastomers can possibly exhibit stress-induced crystallization, the high strength could conceivable result largely from the crystallization process. However, an elastomer based on the non-crystallizable polyoxypropylene glycol, but other-

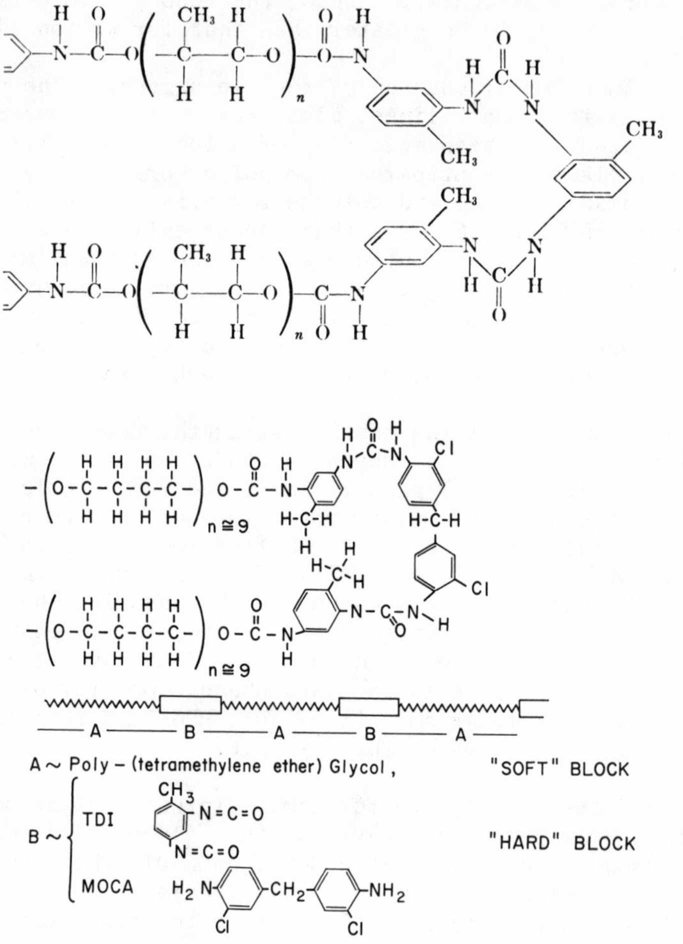

Figure 6   Upper formula is the basic structure in elastomers
prepared from polyoxypropylene glycol, toluene-2,4-
diamine, and toluene-2,4-diisocyanate.   Lower formula
is the basic structure in Adiprene L elastomers.

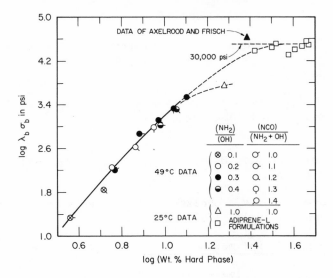

Figure 7   True tensile strength data for several types of segmented
diisocyanate-linked (polyurethane) elastomers.

wise similar to the Adiprene formulations, has been found by
Axelrood and Frisch (28) to have a tensile strength comparable to
those of Adiprene L elastomers, as indicated by the shaded triangle
in Figure 7. Thus, the high strength of the diisocyanate-linked
elastomers must result primarily--if not entirely--from the plastic
domains.

SUMMARY AND CONCLUSIONS

Stress-strain data measured at large deformations from -40 to
60°C on unplasticized and plasticized styrene-butadiene-styrene
triblock polymers, Kraton 101 and Thermolastic 226, are discussed.
It is concluded that the time dependence of the data results
primarily from the plastic deformation and breakdown of polystyrene
domains. Also discussed are the ultimate tensile properties of
three types of elastomeric materials that contain a dispersed plastic
phase: (1) the triblock polymers; (2) styrene-butadiene rubber
vulcanizates filled with colloidal polystyrene spheres; and (3) segmented diisocyanate-linked (polyurethane) elastomers. These materials
owe their strength and toughness to the sizeable energy dissipation
in the vicinity of slowly growing cracks; the dissipation results
primarily from the plastic deformation and ductile rupture of the
plastic domains. To obtain high strength in a non-crystallizable
elastomer over a broad temperature range, it is sufficient to have
about 25% of colloidal plastic particles, uniformly dispersed in a
rubbery network; presumably the particles must deform plastically
under high stress.

REFERENCES

1.   T.L. Smith, "Strength and Extensibility of Elastomers," Rheology, Vol. 5, F.R. Eirich, Ed., Academic Press, New York, 1969, Chapter 4.

2.   See Block Copolymers (J. Polymer Sci. C, 26), J. Moacanin, G. Holden, and N.W. Tschoegl, Eds., Interscience, New York, 1969.

3.   C.S. Schollenberger, H. Scott, and G.R. Moore, Rubber World 137, 549 (1958).

4.   S.L. Cooper and A.V. Tobolsky, J. Appl. Polymer Sci. 10, 1837 (1966); ibid 11, 1361 (1967).

5.   H.A. Vaughn, J. Polymer Sci. B, 7, 569 (1969).

6.   R.P. Kambour, J. Polymer Sci. B, 7, 573 (1969).

7.   D.G. LeGrand, J. Polymer Sci. B, 7, 579 (1969).

8.   R.W. Rees and D.J. Vaughan, Am. Chem. Soc. Polymer Preprints 6, 296 (1965).

9.   T.C. Ward and A.V. Tobolsky, J. Appl. Polymer Sci. 11, 2403 (1967).

10.   T.L. Smith and R.A. Dickie, Block Copolymers (J. Polymer Sci. C, 26), J. Moacanin, G. Holden, and N.W. Tschoegl, Eds., Interscience, New York, 1969, p. 163.

11.   M. Morton and J.C. Healy, Polymer Modification of Rubbers and Plastics (Applied Polymer Symposia, No. 7), H. Keskkula, Ed., Interscience, New York, 1968, p. 155.

12.   M.Morton, J.C. Healy, and R.L. Denecour, Proc. Intern. Rubber Conf. 1967, Maclaren and Sons, London, p. 175.

13.   C.W. Childers and G. Kraus, Rubber Chem. Technol. 40, 1183 (1967).

14.   J.F. Beecher, L. Marker, R.D. Bradford, and S.L. Aggarwal, Block Copolymers (J. Polymer Sci. C, 26), J. Moacanin, G. Holden, and N.W. Tschoegl, Eds., Interscience, New York, 1969, p. 117.

15.   R.H. Valentine, J.D. Ferry, T. Homma, and K. Ninomiya, J. Polymer Sci. A-2, 6, 479 (1968).

16.   J.C. Halpin and F. Bueche, J. Applied Phys. 35, 3142 (1964).

17.   R.P. Kambour, J. Polymer Sci. A-2, 4159 (1964).

18.   G. Holden, E.T. Bishop, and N.R. Legge, Proc. Intern. Rubber
       Conf. 1967, Maclaren and Sons, London, p. 287. Also see
       Block Copolymers (J. Polymer Sci. C, 26), J. Moacanin, G.
       Holden, and N.W. Tschoegl, Eds., Interscience, New York, 1969,
       p. 37.

19.   H. Hendus, K.H. Illers, and E. Ropte, Kolloid-Z, 216-217,
       110 (1967).

20.   N.W. Tschoegl et al, unpublished results.

21.   J.C. Halpin, Rubber Chem. Technol. (Rubber Reviews) 38,
       1107 (1965).

22.   H.W. Greensmith, J. Applied Polymer Sci. 8, 1113 (1964).

23.   A.G. Thomas, The Physical Basis of Yield and Fracture:
       Conference Proceedings, The Institute of Physics and The
       Physical Society, London, 1966, p. 134.

24.   M. Morton, unpublished results.

25.   T.L. Smith, unpublished results.

26.   A.J. Havlik and T.L. Smith, J. Polymer Sci. A, 2, 539 (1964).

27.   Adiprene Bulletin No. 11, Adiprene L-213 Urethane Rubber,
       Elastomer Chemicals Dept., E.I. duPont de Nemours & Co.,
       Wilmington, Del., 1966. Data taken from Table VI.

28.   S.L. Axelrood and K.C. Frisch, Rubber Age 88, 465 (1960).
       Also see, J.H. Saunders and K.C. Frisch, Polyurethanes:
       Chemistry and Technology, Part I, Interscience Publishers,
       New York, 1962, p. 311.

AN ALPHA-METHYLSTYRENE ISOPRENE BLOCK COPOLYMER

Gabriel Karoly*

Union Carbide Corporation, Chemicals & Plastics

Box 670 Bound Brook, N. J. 08805

INTRODUCTION

Considerable interest has been generated during the last few years by the development of styrene-butadiene-styrene (SBS) block copolymers. These polymers were found to be two phase systems, with two glass transition temperatures and with properties closely resembling those of crosslinked filled elastomers. The two glass transition temperatures are almost identical with those of the individual blocks, i.e., -80 and +90° C. Above the glass transition temperature of the hard segments, block polymers lose their mechanical properties. This work discusses an attempt to replace the polystyrene with poly-α-methylstyrene in these block copolymers. The literature reports different glass transition temperatures for poly-α-methylstyrene. Values as high as 184°C were measured in our laboratories. The purpose of this work is to utilize this high glass transition temperature to obtain a thermoplastic elastomer with a high use temperature. According to Holden[1,2] SBS block copolymers can be prepared by living anionic polymerization. Secondary butyl lithium is a suitable initiator for the preparation of the living styrene block, which in turn, initiates the polymerization of the butadiene after this second monomer is introduced. The addition of more styrene to the living polymer completes the preparation. An alternate

*Present address: M&T Chemicals Inc., P. O. Box 1104, Rahway, New Jersey 07065

153

process utilizes the fact that dienes are more reactive
toward anions than styrene. The styrene polymerization
stops short of completion when a diene is added midway
and continues only when the diene is consumed. A third
method is based on coupling. Here again, styrene is
polymerized first, followed by diene polymerization until a living SB-1/2 polymer is prepared. The coupling
of two SB-1/2 polymers then gives the desired SBS
structure.

All of these procedures are based on a living anionic polymerization of styrene before that of the
diene. It is a fact that the counterion to the living
anion has to be lithium, and that the lithium-carbon
bond cannot be further polarized by the introduction
of activating agents, the reason being that in the
presence of even small (0.5 vol. %) ethers[3], or if the
counterion is not lithium but another alkali metal,
predominantly 1,2 polybutadiene and 3,4 polyisoprene
is formed instead of the desired 1,4 adducts. It is
well known that 1,2 polybutadiene and 3,4 polyisoprene
have undesirable properties as elastomers. Their glass
transition temperature is too high, their elongation
too low and they are extremely tacky.

Alkyl lithium compounds in nonpolar solvents are
present as ion pairs, and as such are very slow in polymerizing α-MST. Since the solvation of the counterion
has an undesirable effect on the diene polymerization,
a procedure based on a moninitiator which polymerizes
the hard segment first is unsuitable for the preparation of α-MST-diene-α-MST block copolymers. A new procedure had to be developed where the diene segment is
prepared first with a diinitiator which is solvated
and thereby activated only after the completion of the
diene polymerization. Dilithio-1,4-dimethyl-1,4-di-
phenyl-butane (the dianion from α-MST dimer) was found
to be an excellent initiator, and hexamethylphosphoramide (HMPA) an excellent activator for the polymerization of α-MST.

Materials

Isoprene, practical grade, was distilled directly
into the reactor. α-Methylstyrene research grade,
vacuum distilled at 40°C and 7 mm before polymerization.
Toluene was refluxed with sodium dispersion, and distilled at atmospheric pressure. Lithium was obtained
in a 50% dispersion in AMSCO solvent and used as is.
Hexamethylphosphoramide was vacuum fractionated at 80°C

and 1 mm pressure just prior to being used. Argon gas had a purity of 99.996% and was used as is.

Preparation of the Catalyst

A dry (125 ml) Erlenmeyer flask already containing a magnetic stirrer is charged with 50 ml toluene, 20 ml α-methylstyrene, 1.5 g of a lithium dispersion and sealed with a rubber serum stopper. At room temperature and under vigorous agitation, 6 ml of tetrahydrofuran is added through a hypodermic needle to the system. The catalyst is formed within a few minutes and is usable as is.

Polymerization

Apparatus: 1 liter resin flask equipped with a thermometer, stirrer, Dry Ice reflux condenser, argon inlet and serum stopper. All glassware is dried overnight at 200°C in vacuum, assembled while still hot under argon, and flamed prior to polymerization. The reactor is charged with 500 ml of toluene, 150 ml of isoprene, 70 ml of α-MST and heated to 60°C where 2 ml of catalyst solution is added slowly. A very slight color appears indicating that the system is pure and free of contaminants. A total 6 ml of catalyst is charged. The isoprene starts to polymerize and the temperature rises slowly to 70°C. The water bath used to heat the unit to 60°C is removed and replaced with a wet ice/alcohol bath (-12°C). The solution becomes more and more viscous, as the temperature rises to 90°C.

As soon as all of the isoprene is consumed, which takes about 30 minutes, the temperature starts to drop. It takes a total of one hour to cool the flask to 10°C, at which point 2 ml of HMPA is added. There is an immediate color change from slight yellow to a light orange. This light orange color darkens gradually and in about a minute it turns dark purple, indicating the polymerization of α-MST which takes only a few minutes. Upon the addition of the HMPA the temperature drops first to about 0°C and then rises again to 15°C. The drop is due to the sudden decrease of the viscosity which promotes heat transfer, the rise to the heat of polymerization. The reaction is quenched at 10°C with CO_2 after 15 minutes. The polymer is precipitated with isopropyl alcohol in a Waring Blendor, washed four times to remove all toluene and unreacted α-MST,

stabilized with N-phenyl-β-naphthylamine, and dried in
vacuum at 80°C.

Analytical

A ten per cent solution of the block copolymer in
carbon tetrachloride was analyzed on a 60 Mhz Varian nuclear magnetic resonance spectrometer. The poly-α-methylstyrene content was determined on the basis of aromatic
proton (∿7.0 ppm δ) absorption. The polyisoprene center
block of the copolymer was regarded as a polyisoprene
polymer for which correlation between chemical shift and
structure is known. The absorption in the aliphatic
region was corrected for the aliphatic protons of poly-α-
methylstyrene. Since no peaks appeared around 5.3 ppm
(δ) it was concluded that little or no 1-2 polyisoprene
was formed during the polymerization. By knowing the
chemical shifts for the various groups, the areas under
the peaks, and the molecular weight of the monomers, the
system became fully characterized by a set of simple
first order equations. No attempts were made to resolve
the cis and trans distribution of the 1,4 polyisoprene.

RESULTS AND DISCUSSION

Initiator

The electron transfer reaction between α-MST and
sodium or potassium is known[4,5,6]. It has been shown
that the radical anion dimerizes, and forms a dimer dianion. In the presence of an excess of α-methylstyrene
the dianions could react further to form higher oligo-
mers. Polymerization of α-methylstyrene is prevented
by keeping the temperature close to the ceiling temperature of poly-α-methylstyrene, which is ∿60°C.

Polymerization

The polymerization has to be carried out in two
steps; first the diene segment has to be formed, then
α-MST has to be polymerized at both ends. It is of
paramount importance that once the polymerization has
started, the system should be free from all terminating
impurities. Premature termination can lead either to
I or to AI type polymers where I represents the elastomeric polyisoprene segment and A the thermoplastic

poly-α-MST segment. Polyisoprene homopolymer is incompatible with the AIA block copolymer and while the AI blocks are compatible with the AIA blocks, they lower considerably both the tensile properties and the glass transition temperature. Effective catalyst concentrations are of the order of 10^{-3} - 10^{-4} moles/liter, which means that impurities cannot be tolerated above the 10^{-5} moles/liters concentration. Preparing and transferring materials at these purity levels is extremely difficult, if not impossible, unless micro quantities are involved in a vacuum line.

The initial charge to the reactor includes toluene diluent, α-MST, and isoprene. Toluene is an ideal solvent for the reaction. It is a good solvent both for the catalyst and for the polymer; it does not chain transfer with lithium alkyls; it boils above the polymerization temperature of isoprene, and it can be cooled without freezing. The α-MST does not interfere with the polymerization of isoprene, and therefore, it can be charged prior to the polymerization. The addition of the catalyst has to be very slow at first. The reaction between the catalyst anion and the impurities is instantaneous. Slow addition and good stirring assures essentially complete deactivation of the first portions of the catalyst, and perfect purification of the system. Once the system is free from even traces of impurities the rest of the catalyst is added.

The polymerization takes about 30 minutes if the temperature is permitted to rise to 90°C, longer if it is kept lower. Once all of the isoprene is consumed the solution contains about 20% polyisoprene in toluene. The viscosity of the solution is, however, many fold higher than one would expect from the molecular weight and the concentration. The reason for this is that the living ends are associated making the apparent molecular weight extremely high. Once a small amount of HMPA is introduced, it immediately solvates the lithium counterion. This solvation breaks up the associations of the living ends causing a drastic decrease in the viscosity of the solution. HMPA promoted anions, just like the unpromoted ones, are much more reactive toward dienes than toward α-MST. Therefore, no α-MST polymerization takes place until the system is completely free of dienes. The polymerization of isoprene is quantitative and yields of practically 100% have been obtained. The polymerization of dienes in the presence of HMPA is extremely rapid, but leads, as already indicated, to an undesirable product, namely 1,2 polybutadiene, or 3,4 polyisoprene. Therefore, the introduction of HMPA should be postponed until the diene content is depleted. The polymerization of α-methylstyrene can be followed vis-ually quite readily. The isoprene anion is straw yellow, the α-MST anion dark purple. If the system is completely free of isoprene the color change is instantaneous. If traces of isoprene are still present when HMPA is added, the color change will be gradual.

The anionic polymerization of α-MST is governed by a monomer-polymer equilibrium. The equilibrium monomer concentration is given by the following formula:

$$\ln[M] = \frac{\Delta H - T\Delta S}{RT}^{7}$$

The conversion is a function of this equilibrium monomer concentration and the initial monomer concentration. Due to the low heat of polymerization ($\Delta H \sim 7$ kcal/mole), it is necessary to operate at low temperatures. Satis-factory conversions were obtained between 0 - 10°C at the concentrations used in this work. Accordingly, the conversion depends on the temperature of termination alone and is not influenced by temperature floctuations prior to termination. It was found, however, that a secondary irreversible termination reaction takes place between the living anion and HMPA at or above 30°C. Therefore, it is of great importance that the tempera-ture should not reach 30°C once HMPA is introduced into the system.

The poly-α-MST living ends can be terminated with alcohols, carbon dioxide, epoxides, etc. The means of termination and thereby the nature of the terminal group, was not found to be of importance. Generally, carbon dioxide was used for that purpose.

The polymer composition was determined by NMR. In-formation was obtained not only about the composition of the polymer (wt. % α-MST) but also about the struc-tural distribution of the polyisoprene block. Isoprene, when polymerized anionically, can give a 1,2, a 3,4 and two 1,4 (cis and trans) structures. No attempts were made to resolve the cis and trans distribution. The 1,2 content was considerably less than 1% in all samples analyzed.

The samples were measured for reduced specific vis-cosity at 25°C in toluene. Samples having a reduced

specific viscosity between 0.8 - 1.2 dl/g. (C = 2g/l)
showed optimum physical properties. No attempts were
made to convert the viscosity measurements into molecular weight. K and α values are known from the literature for poly-α-MST and for polyisoprenes in toluene at
20°C. The polyisoprene viscosity molecular weight relationship depends on the isomer distribution. No data
could be found for an isomer distribution even approxi-
mating ours. It is estimated that the samples having
the best physical properties have a molecular weight in
excess of 100,000.

Physical Properties

Table I shows a comparison between an α-MST-isoprene-
α-MST block copolymer and a ST butadiene-ST block copolymer having very similar chemical composition and
molecular weights. The glass transition temperatures
given were determined on an Instron Tester,[a] and by torsion pendulum measurements. On both instruments the
lower Tg associated with the long polyisoprene segment
was sharp, while the upper Tg belonging to the shorter
poly-α-MST segments was not. Therefore, the polymers
were characterized and compared on the arbitrary T_2
bases. T_2 is simply the temperature where the modulus
drops to 100 psi. Above this temperature the polymer
has no mechanical integrity.

Polymers containing both higher and lower α-MST concentrations were also prepared. At low α-MST concentrations the shorter α-MST segments are not fully capable of forming a network and therefore, the polymers
have low modulus, low tensile strength and low Tg.
Above 30-35% α-MST content, samples showed irrecoverable
deformation under load. Therefore, there is only a
narrow range of overall composition where these block
copolymers behave like crosslinked elastomers.

Rheological studies were carried out on a capillary
and on a rheogoniometer. Figure 1 shows the melt viscosity as a function of shear rate for the polymer
described in Table I. These data indicate that the melt
viscosities are very high, and that raising the temperature by 30°C has only a minor effect. The extrudates
showed severe "melt fractures" even at the higher temperature, and at shear rates as low as 1.8 sec.$^{-1}$.

(a) Measuring the modulus as a function of temperature.

TABLE I

COMPARISON BETWEEN α-MST AND STYRENE
ABA BLOCK COPOLOYMERS

Composition	α-MST-[e] Isoprene α-MST	ST-BD-ST[a]
Wt. % Hard block[b]	26	25
Reduced Specific Viscosity	1.2	1-1.5
Properties		
Tg °C[c]	-45	-80
T$_2$ °C[d]	+160	+90
Tensile Modulus psi	1000	700
Tensile Strength psi	2400	3000
100% Modulus psi	230	200
Elongation %	1100	1000

a) Shell Thermolastic 125, according to Shell Chemical Company bulletin.

b) Styrene or α-MST.

c) Glass transition temperature.

d) Temperature at which the tensile modulus is 100 psi.

e) Distribution of the polyisoprene block.

1,4 polyisoprene	74%
3,4 "	25%
1,2 "	<1%

Reduced specific viscosities were measured on samples prior and after extension on the capillary viscosimeter. The results showed no sign of polymer degradation.

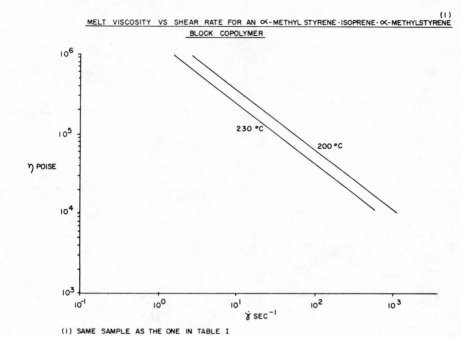

MELT VISCOSITY VS SHEAR RATE FOR AN α-METHYL STYRENE-ISOPRENE-α-METHYLSTYRENE BLOCK COPOLYMER

(1) SAME SAMPLE AS THE ONE IN TABLE I

REFERENCES

1. U. S. Patent 3,265,765.

2. Dutch Patent 660,3846.

3. I. Kuntz and A. Gerber, J. of Polymer Science 42, 303.

4. C. L. Lee, J. Smid, M. Szwarc, J. of Physical Chem., 66, 904.

5. R. L. Williams, and D. H. Richards, Chemical Communications 1967, 414.

6. C. F. Frank et al., J. Org. Chem., 26, 307.

7. F. S. Dainton and K. J. Ivin, Quar. Rev., 12, 61 (1958).

SYNTHESIS AND THERMAL TRANSITION PROPERTIES

OF STYRENE - ETHYLENE OXIDE BLOCK COPOLYMERS

James J. O'Malley, R. G. Crystal, P.F. Erhardt

Research Laboratories, Xerox Corporation,

Rochester, New York   14603

INTRODUCTION

Block copolymers are composed of two or more monomers which are segregated into blocks along the polymer
chain. The copolymers described in this paper have one
block composed of polystyrene (PS) and one or two blocks
composed of poly (ethylene oxide) (PEO). These materials were first synthesized by Richards and Szwarc (1)
and their investigation of the collodial properties of
the copolymers in solution generated considerable interest in the PS-PEO system. Subsequent studies by
Skoulios et.al. (2-5) showed that phase separation of
the copolymer blocks occurred when concentrated solutions of the copolymers were prepared in a solvent preferential for one of the components. The mesomorphous
phases were cylindrical, lamellar and spherical in
structure and were interconvertible by varying the concentration of the polymer solution. Sadron (6,7) was
able to preserve the various mesophases indefinitely
by using polymerizable monomers as the preferential
solvents and polymerizing the solvent. A recent study
by Kovacs and Lotz (8,9) demonstrated that single crystals of the block copolymers can be grown from dilute
solution when the solvent is preferential for PS and
that they are composed of PEO folded chain lamellae
sandwiched between surface layers of amorphous PS.

In our studies on the styrene-ethylene oxide copolymers we have focused on the morphology, rheology
and thermal properties in solvent-free systems. The

present paper describes the synthesis of the PS-PEO
and PEO-PS-PEO block copolymers and discusses their
thermal transition properties. The effects of copolymer structure and composition, as well as those arising
from solvent history, on the transition temperatures
and the crystallinity of the copolymers are discussed.
Subsequent papers (10,11) will describe, in detail, the
morphologies of these block copolymers and their rheological properties.

EXPERIMENTAL

The block copolymers were synthesized by anionic
polymerization in an all-glass high vacuum system in
order to minimize impurities which cause premature termination of the polymerization reaction. Tetrahydrofuran, which was prepurified by distilling from potassium metal and given a final distillation from sodium
benzophenone dianion, was used as the polymerization
solvent. Styrene monomer was freed from inhibitor and
dried by two distillations from fresh calcium hydride.
The final distillation was made into a calibrated am-
poule and the monomer was diluted with dry THF. Ethylene oxide was purified by two distillations from fresh
calcium hydride at -30°C and it was added to the polymerization reactor undiluted with solvent.

Cumyl potassium and dipotassium α-methyl styrene
tetramer dianion were used as the polymerization initiators for the preparation of PS-PEO and PEO-PS-PEO
block copolymers respectively. The former was prepared
in THF from methylcumylether and sodium-potassium alloy
according to the method of Ziegler (12). The latter
initiator was prepared by reacting a 0.4 molar solution
of α-methyl styrene in THF with a potassium mirror at
room temperature for approximately one hour. The concentrations of carbanions in both initiator solutions
were determined spectroscopically and by titration of
a known volume of the initiator solution (previously
reacted with a few drops of water) with 0.05 molar HCl.

Copolymer Synthesis

The block copolymers were prepared by a two stage
process. In the first stage, living polystyrene was
synthesized by adding a THF solution of styrene monomer
to a well-stirred solution of initiator in THF. The
monomer solution was added through a capillary tube so

that it contacted the initiator solution as a fine spray
or mist. The polymerization was allowed to proceed for
0.5 hours at 0°C. The initiation and polymerization of
ethylene oxide by living polystyrene was the second
stage of the process. The monomer was added to the
living polymer solution at -78°C and a rapid discolor-
ation of the red living polystyrene solution was observed. The polymerization solution was then stirred
at room temperature for 24 - 36 hours to complete the
polymerization. The polymerizations were terminated by
addition of a drop of glacial acetic acid and the polymers were isolated by precipitation in hexane. Complete
conversion of monomer to polymer was found in each experiment.

Purification of the block copolymers was carried
out in order to free them from unwanted homopolymers.
Polystyrene homopolymer was removed by fractionally
precipitating the block copolymer and any homo-poly
(ethylene oxide) from a benzene solution with diethyl
ether (Method A) or by extracting the block copolymer
with warm (45°C) cyclohexane under a nitrogen atmosphere (Method B). Homo-PEO was removed by extracting
the mixture with either alcohol or water, depending on
whether the major component of the block copolymers
was poly(ethylene oxide) or polystyrene respectively.
Approximately 80% of the original polymer sample was
recovered after purification. Practically all losses
were incurred in the alcohol or water extraction step
and some block copolymer was undoubtedly lost during
this operation since infrared spectra indicated polystyrene was present in the extracted polymer.

Copolymer Characterization

The composition of the copolymers was determined by
elemental analysis and ultraviolet spectroscopy using
the styrene absorption band at 260 nm. The results of
these independent techniques were in excellent agreement with one another.

Number average molecular weights of the block copolymers were determined by membrane osmometry at 37°C
in chlorobenzene. Agreement between the experimentally
determined molecular weight and the molecular weight
calculated from the anionic polymerization parameters
was good. These data are shown in Table 1 and they are
indicative of the purity of our block copolymers.

TABLE I

Block Copolymer Composition and Molecular Weight

Sample	Structure	% PS		$\overline{M}_n \times 10^{-3}$	
		Wt.	Mole	Expt.	Calc.
25	PEO-PS-PEO	23.5	11.4	8.5-5.1-8.5	9.3-5.0-9.3
29	"	38.5	21.0	8.4-10.5-8.4	9.3-10.0-9.3
31	"	60.2	38.8	9.8-29.8-9.8	11.2-32.7-11.2
33	"	70.3	49.8	9.1-43.2-9.1	10.0-47.0-10.0
54	PS-PEO	70.0	49.5	24.3-10.4	20.4-10.2
57	"	66.7	45.6	46.8-23.4	40.2-20.1
61	"	17.8	8.3	9.6-44.2	10.0-40.0
63	"	19.6	9.6	21.2-87.2	20.0-80.0
65	"	28.2	14.1	16.0-40.7	20.0-40.0
84	"	39.3	21.4	13.2-20.3	10.0-20.0

A Perkin Elmer Differential Scanning Calorimeter Model 1B was used for the thermal characterization studies. For the melting temperature measurements sample sizes in the range of 4-8 mg were used and the temperature scan speed was 10°/min. The melting points recorded at 0.625°/min. were identical to those recorded at 10°/min. For the glass transition measurements, sample sizes ranged from 20-40 mg and a scan speed of 20°/min. was used. The samples were cycled through the glass transition temperature region five times and an average value of Tg was calculated. A thermogram of one of the block copolymers is shown in Figure 1 to illustrate how the transition temperature data were extracted from the thermograms.

RESULTS AND DISCUSSION

The anionic polymerization of the ethylene oxide portion of the block copolymer is conventionally carried out at 60 - 75°C over a 3 - 7 day period using either sodium or potassium counterions (1,13,14). Frequently, however, in the early stage of the reaction the vapor pressure at these temperatures is sufficient to shatter a sealed thick-walled Carius tube. During our initial experiments a large increase in the viscosity of the polymerization solutions was observed when the solutions

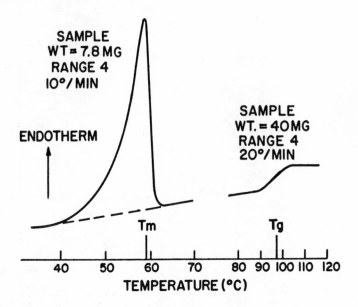

Fig. 1. Thermogram of a PS-PEO block copolymer showing the melting and glass transition temperatures.

were stirred overnight at 25°C before heating to 60°C. This procedure also eliminated the explosion problem at the higher temperature. These observations suggested that the ethylene oxide polymerization was proceeding at a reasonable rate at 25°C and this was confirmed in an experiment which monitored the pressure drop in the reaction vessel as a function of time. The pressure decreased rapidly during the first ten hours and after 16 hours leveled off at a constant value (Fig. 2). The conversion of monomer to polymer was found to be 95% after 16 hours at 25°C. The import of this finding was that copolymers could be synthesized much more rapidly than previously reported and at temperatures and pressures that did not require special heavy walled high pressure glassware. The relatively mild reaction conditions also facilitated larger scale preparations of the block copolymers.

The rate of ethylene oxide polymerization was also found to depend on the counterion and it increased as the electropositivity of the alkali metal increased. With lithium as counterion less than 50% conversion of monomer to polymer was achieved after 72 hours at 25°C

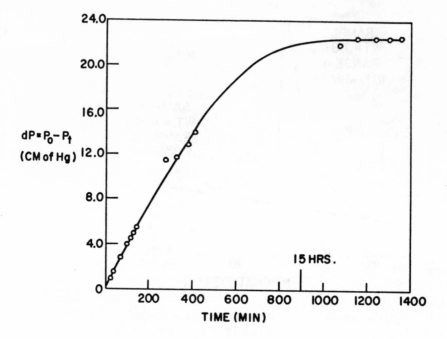

Fig. 2.   The decrease in pressure accompanying the
    cumylpotassium initiated anionic polymerization
    of ethylene oxide in THF at 25°C.

whereas with potassium counterion conversions reached
95% in 16 hours. Other reports (15,16) confirm this
trend in the relative reactivity of the alkali metal
alkoxide ion pairs toward epoxide ring opening in THF.

     Thermograms of the styrene-ethylene oxide block co-
polymers listed in Table 1 show thermal transitions
corresponding to the melting of crystalline PEO and the
glass transition of PS and provide convincing evidence
for phase separation in both the two and three block
systems. In the PEO-PS-PEO series, the molecular weight
of the terminal PEO blocks was held constant at approximately 9,000 g/mole and the copolymer composition was
varied by changing the molecular weight of the central
PS block. Films of these copolymers were prepared by
solution casting at room temperature from chloroform,
a good solvent for both components, and their melting
points are shown in Figure 3. The results indicated the
PEO blocks crystallize even in the presence of 70 wt.%
amorphous PS but that the melting points of the PEO
blocks decrease as the amount of PS in the copolymers

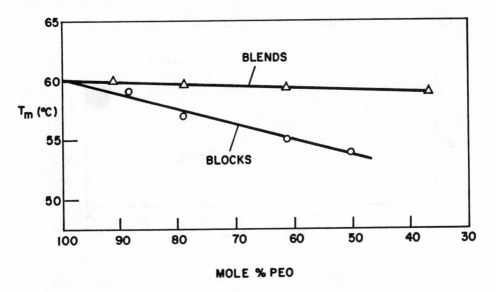

Fig. 3. Melting points of chloroform cast films of PEO-
    PS-PEO block copolymers and polyblends of PS(MW=20,000)
    and PEO(MW=10,000).

increases. Included in Figure 3 are data for a comparable
set of polyblend samples of PS(MW=20,000) and PEO(MW=
10,000). No depression in the PEO melting point is
observed throughout the polyblend composition range and
the data indicate the polymeric components phase separate
into domains of sufficient size to preserve the properties of the individual homopolymer components. Macroscopic phase separation, as occurs in the polyblends,
is not possible in the block copolymers because the incompatible components are linked through covalent bonds.

Melting point data for PEO-PS-PEO copolymer films
cast from preferential solvents and by cooling melts
from 115°C at 10°/min. in a nitrogen atmosphere are
shown in Figure 4. The melting points once again decrease
significantly as the PS content of the copolymers increase and, at 50 mole % PS, the melting point of the
PEO block is 8 - 9°C less than homo-PEO of the same
molecular weight. A similar dependence of melting point
on copolymer composition is found with samples in the
PS-PEO series which have the same PEO block length. For

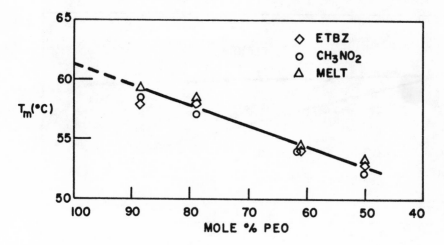

Fig. 4. Melting points of PEO-PS-PEO block copolymer
    films cast from preferential solvents and the melt.

example, samples #57 and #84 have melting points of 56°
and 59°C respectively at a PEO molecular weight of about
20,000 and samples #61 and #65 having melting points of
63° and 61°C respectively at a PEO molecular weight near
40,000. The trend in melting point with composition
appears to be independent of the film preparation method
even though the casting solvents, in particular, have
much different solvating characteristics. In ethylbenzene, the PS portion of the copolymer is preferentially solvated whereas in nitromethane, the preferentially solvated component is PEO. In the dry state, copolymer films cast from these solvents have quite different morphologies (10) and mechanical properties (11)
but it appears that DSC is relatively insensitive to
these differences.

    The melting point depression found with the PEO-PS-
PEO copolymers is of interest since it is expected (17)
to be negligibly small and independent of composition
when the crystallizable block sequence is large (DP≥100)
and the system is at thermodynamic equilibrium. Clearly,
the first condition is satisfied in these samples and,
since the PEO molecular weight is approximately constant
in all the samples, the depression cannot be attributed
to molecular weight effects. Electron microscopy, however, suggests possible reasons for the melting point

depression. Transmission electron micrographs (10) of
all the copolymer samples reveal the basic PEO lamellar
structure but the size and perfection of the lamellae
appear to depend on copolymer composition. As the PS
content in the copolymers increases, the lateral dimensions of the lamellae become smaller and the lamellae
become progressively more distorted. These observations,
coupled with the DSC results, suggest phase separation
of the blocks becomes less complete as the PS content
in the copolymers increases and these imperfections lead
to the observed melting point depression shown in Figures
3 and 4.

    The glass transition temperatures (Tg) of the PS
segments in the block copolymers were measured by DSC
and are shown in Table 2. In each sample the Tg of the
PS block is higher than the melting point of the PEO
blocks. The data for the PEO-PS-PEO series also indicate a progressive increase in Tg as the PS content in
the copolymers increases. There is, however, a con-
current increase in the PS molecular weight in this
series which suggests that the Tg is governed by a
molecular weight dependence (18) rather than by copolymer
composition. The data for PS-PEO samples #54 and #65
lend support to this conclusion since their glass temperatures vary by only 5°C while their PS contents vary
from 70.3 wt.% to 28.2 wt.% respectively. A similar
conclusion was reached during an earlier study of this
system (19) but there is not general agreement on this
point (20).

    In addition to melting and glass transition tempera-
ture data, the thermograms also provide information re-
garding the heats of fusion of the copolymers. This
quantity was determined by measuring the area under the
melting endotherm and converting it to calories per
gram of PEO in the sample. The degree of crystallinity
of the PEO segment was then calculated from the ratio
of the experimentally determined heat of fusion of the
copolymers and the value of 45 calories per gram determined by Mandelkern for the latent heat of fusion
of 100% crystalline PEO (21). The crystallinity data
for the PEO-PS-PEO copolymers shown in Figure 5 were
obtained using this method. The degree of crystallinity
or the weight fraction of PEO in the copolymer which
crystallizes decreases as the copolymer becomes rich in
PS. As was found in the melting point data in Figure 4,
there appears to be essentially no effect of casting
solvent on the copolymer crystallinities. However, the
crystallinity which develops on cooling from the melt
(115°C) at 10°/min. appears to be less than that found

TABLE 2

Glass Transition Temperatures of the Polystyrene
Segment in the Styrene-Ethylene Oxide Copolymers

Polymer	Mole % PS	$\overline{M}_n$ (PS block)	Tg (°C)
ABA25	11.4	5,100	75
ABA29	21.0	10,500	83
ABA31	38.8	29,800	100
ABA33	49.8	43,200	101
AB54	49.5	24,300	96
AB57	45.6	46,800	98
AB65	14.1	16,000	91
AB84	21.4	13,200	88

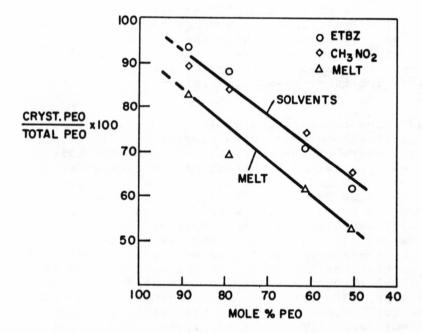

Fig. 5.   The degree of crystallinity of the PEO block
in solvent and melt cast films of the PEO-PS-PEO
block copolymers.

by solvent casting the films. The high viscosity of the melts, chain entanglements, and the presence of a pre-formed glassy PS phase during crystallization of the PEO blocks may account for the lower crystallinity of the films prepared from the melt.

The trend of decreasing PEO crystallinity with increasing PS content is also observed in the PS-PEO copolymers and, with the exception of copolymer #63, the data for the di- and tri-block samples show approximately the same crystallinity-composition dependence. The exception, noted above, has a high molecular weight (87,200 g/mole) PEO block and a lower than expected degree of crystallinity. It has been reported, however, that PEO crystallinity begins to decrease above a molecular weight of 30,000 g/mole (22,23).

The effects of copolymer structure on thermal properties can be evaluated by comparing the copolymers shown in Table 3. These copolymers have essentially the

TABLE 3

Effect of Block Copolymer Structure on
Transition Temperatures and Crystallinity

Measurements	Block Copolymers		
Sample #	33	54	57
Structure	PEO-PS-PEO	PS-PEO	PS-PEO
$\overline{M}_n$ X 10^{-3}	9.1-43.2-9.1	24.3-10.4	46.8-23.4
Mole % PEO	50.2	50.5	54.4
T_g (°C)	101	96	98
T_m (°C)	52[a] 53[b]	53[a] 54[b]	55[a] 56[b]
% PEO Cryst.[c]	66[a] 62[b]	75[a] 65[b]	69[a] 64[b]

a) Films cast from nitromethane
b) Films cast from ethylbenzene
c) Weight fraction of PEO in copolymer which is crystalline X 100, i.e., wt. crystalline PEO/total PEO X 100.

same composition but copolymer #33 is a tri-block and the others are di-block polymers. The data indicate that the thermal parameters are sensitive to the size of the copolymer blocks but are relatively unaffected by their placement in the copolymers.

An interesting observation was made during our experiments on copolymer crystallization from the melt that has a close analogy in the experiments of Price (24) on the crystallization of molten PEO droplets dispersed in silicone oil. When the PS content is less than 40 mole% and the cooling rate from 115°C is 10°/min., the PEO blocks crystallize rapidly near 40°C, which corresponds to an undercooling of 15-20°C. In copolymers with higher PS contents, the PEO blocks exhibit two crystallization exotherms, one near +40°C and the other near -20°C. Similar observations have been recently reported by Lotz and Kovacs using dilatometry (20). In the case of PS-PEO copolymer #57, these crystallization exotherms have approximately the same energy (Fig. 6). The occurrence of two fusion exotherms spaced 60°C apart suggests a difference in nucleation phenomenon at the two temperatures and that homogeneous nucleation may become important at high PS concentrations and when PEO forms the discrete phase. Similarly spaced but energetically unequal fusion exotherms are observed for copolymer #54 (Fig. 7), which has approximately the same composition as copolymer #57 but only one-half its molecular weight. Further evidence is necessary before attributing the differences in Figures 6 and 7 to the molecular weight of the PEO block. However, these observations do suggest that block copolymers may provide convenient and useful materials for studying nucleation phenomena in high polymers.

CONCLUSIONS

Anionic polymerization techniques were used to synthesize block copolymers of polystyrene and poly(ethylene oxide). The use of monofunctional initiators led to copolymers containing two blocks, PS-PEO, and difunctional initiators gave structures with three blocks, PEO-PS-PEO. A brief kinetic study of the ethylene oxide polymerization resulted in a relatively fast, low temperature, low pressure synthesis of the block copolymers and established the relative reactivities of the alkoxide-alkali metal ion pairs toward epoxide ring opening. The rate of ethylene oxide polymerization increased as the electropositivity of the alkali metal counterion increased.

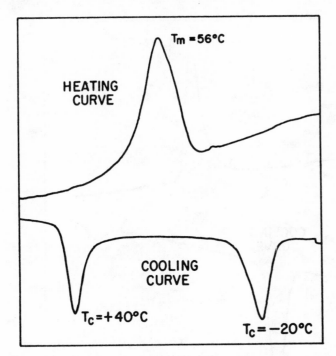

Fig. 6. Heating and cooling thermograms of PS-PEO
copolymer #57. Cooling curve shows two fusion
exotherms on crystallization of the copolymer from
the melt.

DSC was used to characterize the thermal properties
of the block copolymers. For polymers having PEO segments with equivalent molecular weights, the melting
point of the crystalline PEO decreased as the amount of
PS in the copolymers increased. A similar trend was observed when copolymer crystallinity and composition were
compared. The preparation of copolymer films from preferential solvents for the copolymer components did not
change the melting point or the crystallinity of the
copolymers. The variation in melting point with copolymer imperfections (electron microscopy) as the PS
content in the copolymers increased. Films prepared from
the melt had lower crystallinities than those prepared
from solution, but there was no difference in melting
point. The lower crystallinity in melt cast films was
attributed to melt viscosity and chain entanglement
effects.

The effect of copolymer structure on thermal properties was evaluated by comparing PS-PEO and PEO-PS-PEO
copolymers with the same composition. The thermal para-

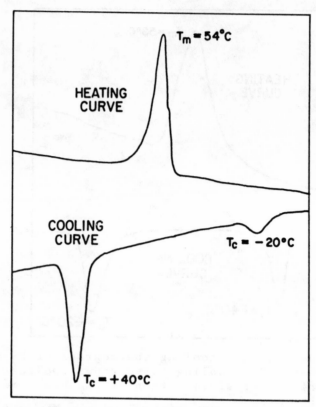

Fig. 7.   Heating and cooling thermograms of PS-PEO co-
     polymer #54. Cooling curve shows two fusion exo-
     therms on crystallization of the copolymer from
     the melt.

meters were sensitive to the size of the copolymer blocks,
but were relatively unaffected by their placement in
the copolymer.

     Finally, observations were made on the fusion (crys-
tallization) temperature of various copolymers as they
were rapidly (10°C/min.) cooled from the melt. At high
PEO compositions a single fusion exotherm near +40°C
was observed whereas at high PS compositions an additional exotherm (of comparable magnitude) was observed at
-20°C. Homogeneous nucleation of PEO probably occurred
at the lower temperature and heterogeneous nucleation
at the higher temperature.

     The authors gratefully acknowledge the assistance
of W. Stauffer, D. Jacobs and G. Sitaramaiah in obtaining the calorimetric and molecular weight data.

References

1. D. H. Richards and M. Szwarc, Trans. Faraday Soc., 56, 1644 (1959).

2. A. Skoulios, G. Finaz and J. Parrod, Compt. Rend., 251, 739 (1960).

3. A. Skoulios and G. Finaz, ibid, 252, 346 (1961).

4. A. Skoulios and G. Finaz, J. Chim. Phys., 59, 473 (1962).

5. E. Franta, A. Skoulios, P. Rempp and H. Benoit, Makromol. Chem., 87, 271 (1965).

6. G. Finaz, A. Skoulios and C. Sadron, Compt. Rend., 253, 265 (1961).

7. Ch. Safron, Pure and Appl. Chem., 4, 347 (1962).

8. A. J. Kovacs and B. Lotz, Kolloid-Z.Z. Polymere, 209, 97 (1966).

9. B. Lotz. A. J. Kovacs, G. A. Bassett and A. Keller, ibid, 209, 115 (1966).

10. R. G. Crystal, J. J. O'Malley and P. F. Erhardt, Block Polymers, Plenum Press, New York, N. Y. (1970), p. 179.

11. P. F. Erhardt, J. J. O'Malley and R. G. Crystal, ibid., p. 195.

12. K. Ziegler and H. Dislich, Ber., 90, 1107 (1957).

13. G. Finaz, P. Rempp and J. Parrod, Bull. Soc. Chim. Fr., 262 (1962).

14. M. Baer, J. Polymer Sci., A2, 417 (1964).

15. A. Rembaum, J. Moacanin and E. Cuddihy, J. Polymer Sci., C4, 529 (1964).

16. V. N. Zgonnik, L. A. Shibaev and N. I. Nickolaev, Paper presented at the IUPAC meeting, Budapest, Hungary 1969.

17. L. Mandelkern, Crystallization of Polymers, McGraw-Hill Inc., New York, N. Y. (1964), pp. 78-79.

18.   T. G. Fox and P. J. Flory, J. Appl. Phys., $\underline{21}$, 581 (1950).

19.   J. J. O'Malley and R. H. Marchessault, Paper presented at the Am. Phys. Soc. Meeting, Philadelphia, Pa., 1969.

20.   B. Lotz and A. J. Kovacs, Paper presented at the Am. Chem. Soc. Meeting, New York, N. Y., 1969.   Polymer Preprints, $\underline{10(2)}$, 820 (1969).

21.   L. Mandelkern, J. Appl. Phys., $\underline{26}$, 443 (1955).

22.   T. M. Connor, B. E. Read and G. Williams, ibid., $\underline{14}$, 74 (1964).

23.   J. A. Faucher, J. V. Koleske, E. R. Santee, Jr., J. J. Stratta and C. W. Wilson III, ibid., $\underline{37}$, 3462 (1966).

24.   F. P. Price, Paper presented at the IUPAC Meeting, Wiesbaden, Germany, 1959.

THE EFFECTS OF PREFERENTIAL SOLVENTS ON THE PHASE SEPARATION

MORPHOLOGY OF STYRENE-ETHYLENE OXIDE BLOCK COPOLYMERS

Richard G. Crystal, Peter F. Erhardt, James J. O'Malley

Xerox Research Laboratories

Rochester, New York 14603

INTRODUCTION

Phase separation in block copolymer systems exhibits rather striking effects on resultant physical properties (1,2). Skoulios et al (3-6) have studied the morphology of block copolymers of poly(ethylene oxide) (PEO) and polystyrene (PS) as well as poly (propylene oxide) and PS, in concentrated solutions using small angle x-ray techniques. By varying both concentration and solvent type, they were able to create several interesting gel structures. As each component in their block systems had markedly different solubilities they were able to explore the effect of dispersing these polymers in preferential solvents. As an example, the more polar PEO was soluble primarily in polar solvents (nitromethane, water, etc.) while the non polar PS segment was soluble primarily in non polar media (ethylbenzene, p-xylene, etc.). By using these preferential solvents, several structures, namely the sphere, the cylinder and the lamella, were identified from small angle x-ray data. More recently, Kovacs et al (7-9) have studied single crystals of PS-PEO AB type blocks and have shown that PEO crystallizes with little interference from the glass forming PS in dilute solution experiments. They also indicate that the resultant PEO crystal lattices are identical to that of the homopolymer.

References to PS-polybutadiene (PB) and PS-polyisoprene (PI) block copolymer morphology have also been reported recently. Aggarwal et al (1) have published electron micrographs of phase separation in PB-PS block copolymers along with mechanical property measurements. Wilkes and Stein (10) have reported light scattering data from Shell Kraton polymers. Electron micrographs by Japanese workers (11,12) suggest marked morphological effects

179

caused by casting PS-PI blocks having varied compositions from preferential solvents.

We have investigated the effects of preferential solvents on the bulk or internal morphology of PS/PEO block polymers having both varied compositions and segment molecular weights. The principal solvents used in this study were nitromethane [preferential to poly(ethylene oxide)], ethylbenzene (preferential to polystyrene) and chloroform (good solvent for both species). Observations described in this paper are restricted to AB copolymers only. PEO-PS blocks were chosen primarily for the marked differences in the properties and behavior of the individual components. While PEO is crystallizable, polar and hydrophilic, PS is glassy, non polar and hydrophobic. Because of their hydrophilic-hydrophobic nature, these materials can, as a first approximation, be regarded as 'macromolecular soaps'. Certain phenomena associated with soaps, such as micellization may indeed show up in solution property studies. It should be kept in mind that because crystalline PEO is not 100% crystalline, bulk polymers should be three phase systems consisting of amorphous and crystalline PEO as well as amorphous PS.

EXPERIMENTAL

A.   Materials

Copolymer formulations were synthesized using ionic polymerization techniques. The preparation, purification and characterization of our materials are described in this volume (13). Table I summarizes characterization of polymers specifically referenced in this study. As a short notation, these polymers will be referred to in the text by their weight percent polystyrene, i.e. the first polymer will be referred to as 19.6% PS.

Table I.   Block Copolymer Sample Characterizations

Type	Wt.% PS	$\overline{Mn}_{Total}$	$\overline{Mn}_{PS}$	$\overline{Mn}_{PEO}$	[η] [dl/g]
AB	19.6	108,000	21,000	87,000	1.13
AB	39.3	33,500	13,200	20,300	0.39
AB	66.7	70,200	46,800	23,400	0.610
AB	70.0	34,800	24,400	10,400	0.405

B. Optical Microscopy

Optical microscopy observations were made on a Leitz Ortho-lux microscope equiped with polarized light, phase contrast and hot stage capabilities. In preferential solvent effect studies, films 5 to 10 microns thick were cast and dried at very slow rates on glass slides from 1% polymer solutions. Films for hot stage studies, however, were prepared by melting polymer granules fol-lowed by pressing the melt between slide and cover glass. Signs of birefringence were determined with the use of a first order red wave plate and polarizing optics.

C. Electron Microscopy

Electron microscopy observations were made on a Jeolco EM 57 single condenser instrument. Both replication and transmission techniques were utilized.

1. Replication Techniques. Bulk morphologies were examined by fracture replication while film surface morphologies were examined by free surface replication. In all cases, one step platinum-carbon replicas were made. Replicated specimens were dissolved away with chloroform, a mutual solvent for both compo-nents.

2. Transmission Techniques. The application of transmission electron microscopy to polymer studies is still a quite virgin field. As a result considerable effort was devoted to technique development for this particular block system. Because the PS and PEO components are hydrophobic and hydrophilic respectively, neither water nor organic solvents could be used as casting and transferral media since one component was always soluble. Castings were performed on both mercury surfaces and fully dried formvar films. In the former method, specimens were picked up on formvar coated grids. The latter method, formvar casting, was avoided as much as possible since uniform spreading was most difficult. Because of extreme sample brittleness, the formvar casting technique could not be avoided with high PS content specimens. Ideally, films approximately 500Å thick were prepared. The formvar substrates, being quite strong, made carbon over-coating unnecessary.

Upon examination directly in the electron microscope little structure could be readily resolved since electron densities of both PS and PEO are quite similar. To produce contrast, the polymer films were preferentially stained with osmium tetroxide, a fixative used extensively by biologists. Osmium tetroxide is a strong oxidizing agent and in biological specimens preferentially attacks certain areas due to the varying chemical constitution inherent in biological materials. The overall objective in preferential staining is to place a relatively heavy atom at

desired locations giving the stained area a dark appearance in
the electron image. In block copolymer applications, osmium
tetroxide has been used successfully in the PS-PB and PS-PI systems
(1,11,12) in which the conjugated butadiene and isoprene double
bonds are readily oxidized. As no such structure existed in PS-
PEO copolymers, it was hoped that the staining technique could be
refined sufficiently so that staining could be terminated after
one component was 'attacked' but before the second component was
affected. Because of PEO solubility in water, film flotation
on aqueous osmium tetroxide was impossible. Instead, films were
exposed to OsO_4 vapor which surprisingly did not degrade the
specimens. At first the fixative was thought to attack the PEO
component first. In reality, the PS regions were preferentially
stained or darkened in the electron image as evidenced by studies
of the individual component homopolymers. We believe that com-
plexing of osmium with the aromatic PS ring occurs before oxi-
dation of PEO, thus placing the heavy osmium atom in PS regions.
By trial and error, optimum contrast was generated in PS-PEO
blocks after four hours staining with the OsO_4 vapor.

RESULTS AND DISCUSSION

As mentioned in the introduction, unique morphologies can be
created when only one block segment is soluble in the casting
solvent. PS-PEO blocks were cast from three primary solvents,
nitromethane (preferential to PEO), chloroform (mutual solvent
for both components) and ethyl benzene (preferential to PS).
Morphologies were then examined as a function of solvent, relative
composition, molecular weight and block type. The optical microscope proved quite useful in examination but because of the importance of submicroscopic structures, the electron microscope was
used extensively.

Both crystallization growth modes and spherulite textures
were observed in the optical microscope. In virtually all cases,
i.e. all block copolymer specimens cast from all solvents, spherulitic or pseudo spherulitic structures were realized. Indeed, a
minimum of 14 different textures were observed. Before going
further, a word about the appearance of the polymer solutions is
worthwhile. In the case of chloroform, all solutions (approximately 1% polymer) were optically clear. In the cases of high
content PS samples in nitromethane and high content PEO in ethyl
benzene, however, the solutions were quite turbid, indicating
that phase separation on a 'colloidal' scale had occurred prior to
casting. As will be expanded on further, the phenomenon had a
marked effect on resultant morphologies.

As observed with polarized optics, film morphologies had
many spherulitic or spherulite like textures, the appearance of

which was dependent on casting solvent, relative component composition and molecular weight. The submicroscopic structures comprising these spherulites as observed by electron microscopy will be discussed in detail.

The components in PS/PEO block copolymers exhibit markedly different properties and are quite incompatible. Striking variations in morphology have been observed due to the high crystallizability of PEO in contrast to the amorphous nature of atactic PS. In the solid state, all block formulations studied cast from all solvents exhibited negative birefringence. Upon melting of the PEO phase, however, the birefringence sign quickly changed to positive and was observed to increase with increasing PS content. Birefringence was visible even at temperatures above 250°C. Casting solvent markedly influenced morphologies of all samples both on a spherulitic and subspherulitic level. In all cases studied, i.e. with all casting solvents and all specimens containing up to 70% PS, spherulitic or pseudo spherulitic growth (crystalline growth from a central nucleating source) was evident. In general, castings from solvents preferential to PS and castings from blocks containing greater than 40% PS resulted in weakly birefringent and more disordered structures.

The following is a summary of primary observations made in our experiments:

High Concentrations of PEO (greater than 60% by weight)

Nitromethane Castings:
The PEO phase has a propensity for spherulitic crystallization, but the insoluble PS phase inhibits 'clean' spherulite formation. The PS phase segregates into both clusters and amorphous lamellae between PEO fibrils. PEO lamellae are undistorted. (See Figures 1,5)

Chloroform Castings:
The PEO phase has a high propensity for spherulitic crystallization (ringed spherulites) and clean spherulitic structures (apparently unhindered by the PS phase) are formed. The PS phase segregates to interlamellar regions forming a glassy 'crust' on PEO fibrils. (See Figures 2,6)

Ethyl Benzene Castings:
Crystallization slightly resembles spherulite growth. Double chain folded PEO lamellae are formed showing little orientation with respect to the growth direction. PS is the matrix material. (See Figures 3,7)

Moderate Concentrations of PEO (40 to 60% by weight)

Nitromethane Castings:
Spherulitic growth with coarse textures are apparent. PS
clusters are predominant but amorphous PS layers are also sandwiched between PEO lamellae. These PEO lamellae are, however,
somewhat distorted.

Chloroform Castings:
Spherulitic growth with coarse texutres along with little
clustering of PS phases are observed. PS apparently coats the
PEO fibrils. Several types of spherulites with varying growth
rates are apparent.

Ethyl Benzene Castings:
Crystal growth is pseudo spherulitic in that growth proceeds
outward from a central nucleating source, but radial orientation
of PEO crystallites does not occur. PEO lamellae are short and
sometimes clustered into spherical domains interpenetrated with
polystyrene.

High Concentrations of PS (greater than 60% by weight)

Nitromethane Castings:
Spherulites are rather ragged and also smaller than the
higher PEO specimens, but exhibit distinct Maltese cross patterns.
PEO fibrils are single lamellae, but are quite short. PS coats
the PEO and also forms discrete phases.

Chloroform Castings:
Spherulites are rough textured but show radial symmetry.
PEO double lamellae are apparent and form the matrix phase. The
higher molecular weight PS being less soluble apparently precipitates from solution first.

Ethyl Benzene Castings:
Very open and faintly birefringent pseudo spherulites are
observed. Spherical entities 2000 to 4000$\overset{\circ}{A}$ in diameter containing
bent, twisted and fragmented PEO double lamellae interpenetrated
with PS are evident. Small fragments of PEO double lamellae are
found in the PS matrix. (See Figure 10)

For primarily illustrative purposes, a somewhat detailed
description of solvent effects on morphology in two blocks, (1),
a fairly high molecular weight specimen containing 19.6% PS
and, (2), a moderate molecular weight specimen containing 70.0% PS

Fig. 1 - 19.6% PS Cast From
Nitromethane, Crossed Polars

Fig. 2 - 19.6% PS Cast From
Chloroform, Crossed Polars

Fig. 3 - 19.6% PS Cast From
Ethyl Benzene, Crossed Polars

Fig. 4 - PEO Homopolymer,
37,000 M.W., Cast From Chloroform, Crossed Polars

will be presented. Effects found in other compositions will be more or less highlighted and, in some instances, described in more detail.

Figures (1-3) are a series of optical micrographs of films cast from our three principal solvents of a 19.6% PS specimen as viewed through crossed polars. In the first micrograph (Fig. 1), the spherulitic structure exhibits a weak Maltese cross pattern and a very rough texture. 'Birdwing' or 'V' shaped entities are evident and oriented along spherulite radii. The insoluble PS phase has apparently altered the free spherulitic growth common to PEO. In the second micrograph (Fig. 2), casting from chloroform, a quite different more perfect ringed spherulitic structure is evident. In this case, both PS and PEO components were soluble in the solvent and thus crystallization of the PEO phase was un-hindered by PS. In the third micrograph (Fig. 3), however, a very weak pseudo-spherulitic structure is evident. The rings in these spherulites were identified as cracks when examined by phase contrast microscopy. For comparison, a fourth micrograph in Figure (4) shows the spherulitic structure of an anionically polymerized 40M molecular weight PEO cast from chloroform.

Fig. 5 - Surface of 19.6% PS Specimen Cast From Nitromethane, One Step Replica Electron Micrograph

Looking at subspherulitic structure, Figures 5, 6, 7 are replication electron micrographs of film surfaces as seen in Figures 1-3. In the first micrograph (Fig. 5), a casting from nitromethane, the classical 'wheat sheath' spherulitic nucleation pattern of the PEO phase is quite evident. On close inspection, however, globules of the insoluble PS phase decorating the fibrillar PS structure are quite evident. The 'birdwing' texture referred to in the discussion of Figure 1 was found to be caused by further wheat sheath nucleation of spherulite growth as the spherulite grows from its central or primary nucleation site. In the next micrograph (Fig. 6), a casting from the mutual solvent chloroform, the 'perfect' fibrillar subspherulitic structure is evident. These fibrils, however, are 'puffy' looking and are approximately 170 $\overset{\circ}{A}$ in diameter compared to 95 $\overset{\circ}{A}$ diameter fibrils in a control PEO specimen cast and crystallized under identical conditions. PS has apparently coated the crystallized PEO fibrils during solvent evaporation. As further evidence of this coating hypothesis, specimens of this block cast from chloroform, when melted below the glass transition of PS, on several occasions recrystallized to the exact same structure on the microscopic hot stage. Unlike conventional crystallization, the specimens became gradually more birefringent with time until they completely crystallized.

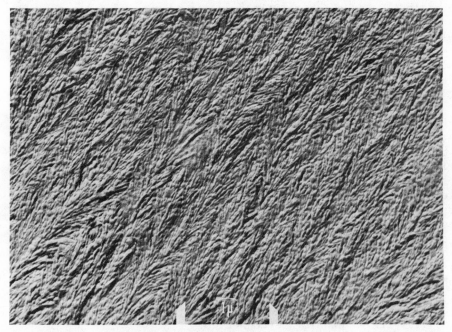

Fig. 6 - Surface of 19.6% PS Specimen Cast From Chloroform, One Step Replica Electron Micrograph

However, crystallizations above the Tg of PS resulted in conventional heterogeneously nucleated spherulitic growth. We assume that below the Tg of PS, the glassy fibril 'crust' locks the PEO melt into the spherulitic skeleton. In the next micrograph (Fig. 7), a casting from ethyl benzene, (preferential to PS), a coarse fibrillar structure is again evident. The PEO apparently crystallizes into fibril groupings or bundles which are oriented at many angles to the radial spherulite direction. This orientation accounts for the rather unordered extinction patterns seen in Figure 3.

Figure 8 is a transmission electron micrograph of a 20% PS film cast from nitromethane. As in Figure 5, the wheat sheath fibrillar structure is quite evident after preferential staining of the PS phase with osmium tetroxide vapor. The dark entity at the wheat sheath center is apparently the heterogeneity responsible for nucleation. PEO fibrils are quite discernable as they splay out of the wheat sheath. Further out, however, their 'trails' become fuzzy due to overlapping with the dark stained PS regions.

Turning attention to a block with higher PS content, Figure 9 is an optical micrograph of a 70% PS film cast from ethyl benzene and viewed between crossed polars. In spite of the radial symmetry of these pseudo spherulitic open structures, no Maltese cross pattern is visible, indicating that PEO lamellae are not radially oriented. While crystallization has proceeded from a central nucleating source, PEO crystallization has apparently ensued before complete solvent evaporation. The same phenomenon has been previously described in this paper in the case of the 19.6% PS specimen cast from ethyl benzene.

Figure 10 is a transmission electron micrograph of the 70% PS block copolymer cast from ethyl benzene and preferentially stained with osmium tetroxide. While PEO double lamellae are clearly identifiable, these lamellae are fragmented, bent and highly distorted. Since crystallization proceeds from a central nucleating source, there must be an interpenetration of PEO lamellae through the bulk structure if crystal nuclei are to be transmitted. Kovacs and coworkers (14) have clearly identified both heterogeneous and homogeneous nucleation in high PS blocks crystallized from the melt. Coupled with calorimetric evidence for incomplete PEO crystallization in this cast specimen (13), the existence of amorphous PEO in this structure must be anticipated. We can conclude, therefore, that this complex structure is composed of crystalline and amorphous PEO and amorphous PS in which the PEO crystallites are somewhat interconnected.

It should be pointed out that in all transmission electron micrographs, the basic PEO fibril element is always present no matter what the morphology. Depending on the relative component

Fig. 7 - Surface of 19.6% PS Specimen Cast From Ethyl Benzene,
One Step Replica Electron Micrograph

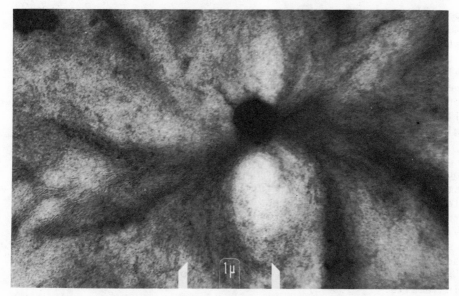

Fig. 8 - 'Wheat Sheath' Pattern in 19.6% Specimen Cast From
Nitromethane, Transmission Electron Micrograph, OsO_4
Staining

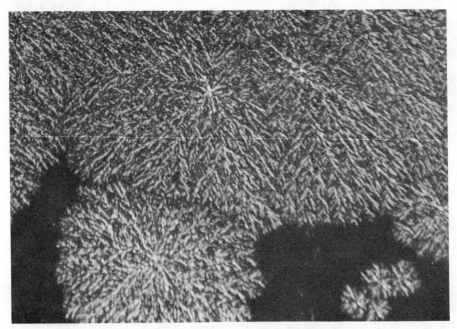

<u>Fig. 9</u> - 70.0% PS Cast From Ethyl Benzene, Crossed Polars

composition and casting solvent conditions, either a 95 Å or 180 Å diameter (thickness) is apparent from the folded chain crystallization of PEO. The 95 Å spacing is clearly evident in crystallizations of a control PEO homopolymer. In general, double PEO lamellae are formed in cases where free PEO crystallization is hampered, i.e. with all specimens cast from ethyl benzene and all chloroform castings having greater than 60% PS in the block polymer. As illustrated in Figure 11, the dimension is related to either a single or double chain folded lamella.

The propensity for PEO to crystallize is extremely great and because of the necessity to form a folded chain structure, PEO crystallization has a major effect on phase separation. To illustrate this point, Figures 10 and 12 are transmission electron micrographs of two specimens cast from ethyl benzene with widely varying PEO content. In Figure 12, a 19.6% PS sample, lamellar and sometimes cylindrical PEO structures are evident. The thicknesses of these PEO regions (light areas due to OsO_4 staining of PS) are 180 Å across, a value comparable to the double folded chain lamella.

In Figure 10, a 70.0% PS specimen larger spherical structures on the order of 3000 to 4000 Å are evident. On close inspection, however, it is clear that these regions, while rich in PEO, are again composed of PEO double lamellae interpenetrated with PS at interlamellar regions. Unlike those of Figure 12, these lamellae,

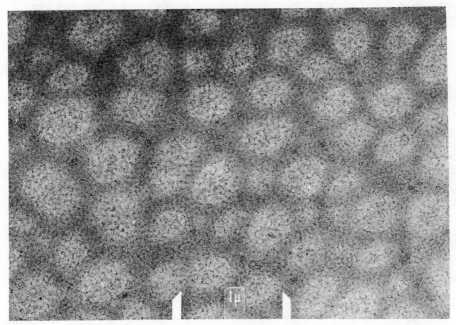

Fig. 10 - 70.0% PS Specimen Cast From Ethyl Benzene, Transmission
Electron Micrograph, OsO_4 Staining

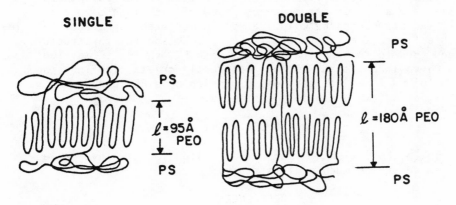

Fig. 11

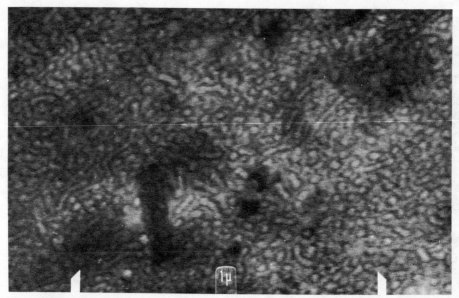

Fig. 12 - 19.6% PS Specimen Cast From Ethyl Benzene, Transmission
Electron Micrograph, OsO_4 Staining

most probably due to the influence of polystyrene, are quite distorted and are inevitably filled with a high degree of defects.

In closing, Figure 10 illustrates two important concepts
generated in our studies. First, the basic phase separation
entity or individual pocket of homopolymer is not the only apparent
structure. Two other morphological entities in increasing order
of size are a larger structure composed of groupings of phase
separated structures as illustrated by the larger spherical entities
in this figure and the spherulite or pseudo spherulite structure.
Second, due to the constraints that each PEO chain is 'married'
to a PS chain and that PEO must apparently crystallize into the
chain folded form, at least one dimension of a PEO phase must be
either one or two chain fold distances.

ACKNOWLEDGEMENT

The authors wish to thank F. J. Walton for his invaluable
assistance during much of this study.

REFERENCES

1. J. F. Beecher, L. Marker, R. D. Bradford and S. L. Aggarwal,
    Polymer Preprints, 8, 2, 1532 (1967).

2. P. F. Erhardt, J. J. O'Malley and R. G. Crystal, Block
    Polymers, Plenum Press, New York (1970), p.195.

3. A. E. Skoulios, G. Tsoulandze and E. Franta, J. Poly. Sci., C-4, 507 (1963).

4. E. Franta, A. Skoulios, P. Rempp, H. Benoit, Die Makromol. Chemie, 87, 271 (1965).

5. C. Sadron, G. Finaz and A. Skoulios, French Patent #1,295,524 (1962), amended (1963).

6. A. Skoulios and G. Finaz, J. de Chimie Physique, 473 (1962).

7. B. Lotz and A. J. Kovacs, Kolloid Z., 209, 2, 97 (1966).

8. B. Lotz, A. J. Kovacs, G. A. Bassett and A. Keller, Ibid, p.115.

9. A. J. Kovacs and J. A. Mason, Ibid, 214, 1, 1 (1966).

10. G. L. Wilkes and R. S. Stein, IUPAC Macromolecular Symposium, Toronto, Ontario, Canada, Sept. 1968 and personal communication.

11. T. Inoue, T. Soen, T. Hashimoto and H. Kwai, IUPAC Macro-molecular Symposium, Toronto, Ontario, Canada, Sept. 1968.

12. T. Inoue, T. Soen and H. Kwai, Reports on Progress in Polymer Physics in Japan, XI, 203 (1968).

13. J. J. O'Malley, R. G. Crystal and P. F. Erhardt, this volume, p.163.

14. A. J. Kovacs, Chemie Industrie-Genie Chemique, 97, 3, 315 (1967).

RHEOLOGICAL PROPERTIES OF STYRENE-ETHYLENE OXIDE BLOCK COPOLYMERS:

TRANSITION AND MELT FLOW BEHAVIOR

Peter F. Erhardt, James J. O'Malley, Richard G. Crystal

Xerox Research Laboratories

Rochester, New York 14603

INTRODUCTION

The block copolymers described herein are long linear sequences of styrene (non-polar, oleophilic, amorphous) units connected by primary chemical bonds to long linear sequences of
ethylene-oxide (polar, hydrophilic, crystallizable) units. Because of the marked chemical dissimilarity of styrene and ethylene
oxide, and the polymeric, long chain nature of the segmental units,
the components are incompatible and phase separation in the melt
is expected and observed. However, because of the primary chemical
bonds tieing together the incompatible chain segments, phase separation is not complete and the rheological behavior of one phase
is strongly affected by that of the other phase. Block copolymers
are unique heterophase systems in this respect and form a distinct
class of materials, the study of which should extend our knowledge
of the manner of flow of mixed phases. The distinguishing feature of the ethylene oxide/styrene block copolymer system is the
sensitivity of the ethylene oxide block to shear. This allows
ethylene oxide blocks to be used as sensitive rheological probes.

There are two complications introduced by the use of the
ethylene oxide/styrene system as a model for the study of the melt
flow properties of block copolymers. One is that the morphology
of the melt is not the same as that of the solid due to the
crystallization of the ethylene oxide at temperatures below 65°C.
In order to elucidate the specific morphology in the melt, observations must be made on the melt itself. Applicable techniques
are hot stage electron microscopy, low angle x-ray scattering and
polarized visible light scattering. Of these methods only the last
has been employed to observations of the melt. The results of that

195

study and its implications on the rheological behavior of ethylene oxide/styrene melts will be considered in a subsequent paper (1). The other complication alluded to above is the known sensitivity of ethylene oxide to thermal or oxidative degradation (2).

Despite the limitations imposed on the rheological studies by the ethylene oxide component, the PEO/PS and PEO/PS/PEO co-polymers are good model systems for block copolymer studies because of the low modulus and low viscosity of the ethylene oxide block segments above 65°C, the melting point of poly(ethylene oxide) and because of the sensitivity of ethylene oxide to shear. This paper will show that many of the principal observations reported on the rheological behavior of SDS, styrene/diene/styrene thermo-plastic elastomers are displayed to a much greater degree by the PEO/PS/PEO systems. The SDS blocks are hard/soft/hard whereas PEO/PS/PEO blocks are soft/hard/soft - where hard and soft are simply relative terms describing the magnitude of the modulus and the viscosity.

A number of experimental investigations relating to the morphological characterization of styrene-ethylene oxide block copolymer gel structures have been reported (3,4,5). Such studies have generally dealt with the influence of preferential solvents in controlling the gel structure of the highly incompatible block components. Although observations have been made of the bire-fringent character of the PS/PEO concentrated solutions and melts, little has been reported concerning the rheological properties of these systems.

For totally amorphous block copolymers, theoretical treat-ments of texture formation based on thermodynamic considerations have been made (6-9). The melt rheology of styrene-butadiene block copolymers has been interpreted by Meier and Arnold (10) and by Holden, Bishop and Legge (11). Such systems are, however, with-out the complicating feature of crystallization by one of the block components. Texture formation studies on ethylene oxide/styrene block copolymers have not been concerned with the melt. The pre-ceeding paper (13) deals with our findings on the block copolymer solids. The purpose of this paper is to relate observations on the melt flow and transition behavior of the AB and ABA styrene-ethylene oxide block copolymers - whose preparation and morpho-logical behavior to specific solvents have been reported.(12,13).

EXPERIMENTAL

Results are reported for those styrene-ethylene oxide block copolymers listed in Table I. All samples were synthesized as previously reported (12). The work-up was generally according to method B (12) in which the samples precipitated from tetra-hydrofuran, THF, by hexane were put into slurry in cyclohexane at

45°C and given final repeated extractions with the liquids indicated in the last column of Table I. The copolymers were subsequently dried, compression molded and tested.

TABLE I

BLOCK COPOLYMER COMPOSITION AND MOLECULAR WEIGHT

Sample	Structure	Wt.% PS	$\overline{M}_n$(total)	$\overline{M}_n$(PEO)	$\overline{M}_n$(PS)	Extractant [a]	
54	PS-PEO	70.0	34.8 K	10.4 K	24.4 K	H_2O	23.4
33	PEO-PS-PEO	70.3	61.3	(2x9.05)	43.2	H_2O	23.4
57	PS-PEO	66.7	70.2	23.4	46.8	H_2O	23.4
31	PEO-PS-PEO	60.2	49.4	(2x9.8)	29.8	MeOH	14.5
84	PS-PEO	39.3	33.5	20.3	13.2	i-PrOH	11.5
29	PEO-PS-PEO	38.5	27.2	(2x8.35)	10.5	Et_2O [b]	7.4
65	PS-PEO	28.2	56.7	40.7	16.0	i-PrOH	11.5
25	PEO-PS-PEO	23.5	22.0	(2x8.5)	5.10	[c]	
63	PS-PEO	19.6	108.4	87.0	21.0	MeOH	14.5
61	PS-PEO	17.8	55.0	45.3	9.73	MeOH	14.5

[a] Immergut and Bandrup, Polymer Handbook, Interscience, New York (1967)
     The numbers indicate the solubility parameter of the extractant.

[b] Et_2O used as precipitant after dissolution in THF of the polymer cleaned
     up according to method B.

[c] Prepared by method A of Reference 8.

Two systems for measurement were employed, a Weissenberg Model
18R rheogoniometer, and a Vibron viscoelastometer, Model DDVII,
modified to measure in the shearing mode, as indicated schematically
in Figure 1a. Data in Figure 1b, modulus as a function of temperature, for one of the block copolymers, is typical of the agreement
found between measurements made on the Vibron and on the rheogoniometer (Rheo).

Fig. 1a - Schematic of the shear deformation chuck used for shear
mode operation of the Vibron Viscoelastometer.

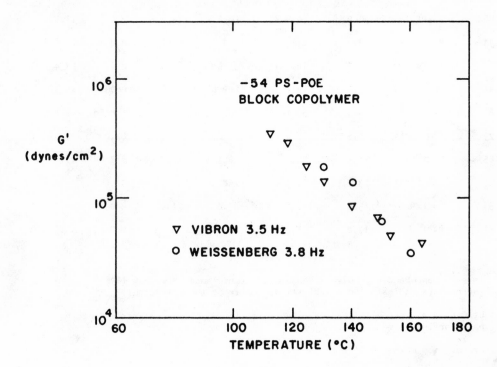

Fig. 1b - Corresponding values of G'(T) obtained on the Vibron (∇)
and on the Weissenberg Rheo (o) for AB block copolymer 54.

RESULTS AND DISCUSSION

Perhaps the most striking features of the styrene-ethylene oxide copolymer melts are: first, the appreciable birefringence even at temperatures above 250°C, and second, the highly elastic character of the melt relative to the component homopolymers. The sign of the birefringence is positive in the direction along the flow lines in the disturbed melt. The rheo-optical behavior of the melts will not be considered further in this paper.

A. Blocks, Blends and Homopolymers

As to melt elasticity, comparison is made with the homopolymers of similar molecular weight and reasonably narrow molecular weight distribution in Figure 2. The dashed lines indicate the temperature dependence of the storage shear modulus, G', for 10.3K monodisperse polystyrene (Pressure Chemical) and for 48K poly(ethylene oxide), prepared by anionic polymerization and having $M_w/M_n \simeq 1.5$. Above 160°C, G' for the AB block copolymer 61 is greater by more than two orders of magnitude than either of the two homopolymer components. An .80/20 PEO/PS blend of these homopolymers was formed by dissolving in chloroform and vacuum evaporating to dryness. Such blends have a $G'(T)$ between that of PS and PEO, but the G' decreases rapidly above 140°C. A gross separation of phases accompanied by coalescence of molten polystyrene globules is observed by hot stage microscopy to occur in this temperature range. The lower modulus observed between 160° and 100°C on cooling is attributable to this irreversible phase separation.

Some difference is observed in $G'(T)$ of the copolymer 61 itself, depending upon the last solvating or extraction treatment received. The sample 61C vacuum dried from chloroform (solubility parameter, $\delta = 9.3$) has a noticeably larger elastic shear modulus component than sample 61B prepared by method B, in which the extracting solvent was methanol ($\delta = 14.5$). The solubility parameter for polystyrene is taken as approximately 9.0 while that for an amorphous melt of $(-CH_2-CH_2-O-)_n$ is calculated to be 8.6 by Small's method (14). Admittedly, the calculation neglects the effects of inter- and intra-molecular hydrogen bonding, which will raise δ by a unit or two, but a value of $\simeq 10.0$, intermediate between those of ethylene oxide ($\delta = 11.5$) and diethylether ($\delta = 7.4$), which serve as crude reference points, appears to be reasonable. On the basis of solubility arguments the higher modulus would be expected for sample 61C where intermolecular entanglements would be more favorable, since at 80% by weight ethylene oxide will always be the continuous phase (15) in the melt.

B. Mechanical Transitions

Figure 3 shows the temperature dependence of the loss tangent,

tan δ, and the storage and loss components of the shear modulus
for the AB copolymer 57. The broad peak in tan δ near 160°C is
evidence of phase separated structure in the melt. The broad peaks
are observable to lesser degrees in all members of the block copolymer series. The appearance of the glass transition region is
evident by the maxima in the loss shear modulus, G'', and tan δ
below 100°C. A corresponding transition in the AB block copolymer
61 was not evident (see Fig. 2). Although, in all cases, two
phases are microscopically apparent, and one always observes the
polystyrene T_g by DSC (12), the magnitude of the mechanical transition corresponding to the T_g of the styrene block is totally
suppressed in some cases.

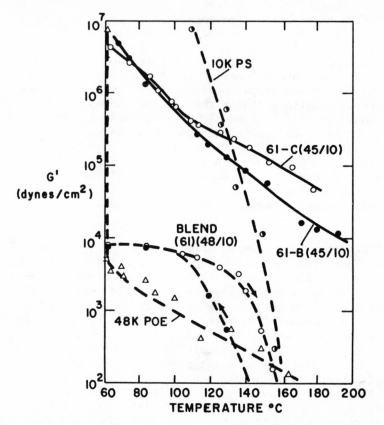

Fig. 2 - G'(T) for the homopolymers PS and PEO, the AB block copolymer 61 and a solution blend of the homopolymers having the
composition of copolymer 61. Numbers in parentheses indicate the
M_n in the order (PEO/PS). Suffix B indicates preparation of the
copolymer by method B; suffix C represents preparation by vacuum
evaporation from chloroform solution.

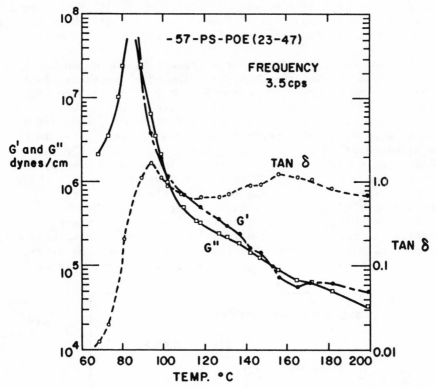

Fig. 3 - G', G'' and tan δ as a function of temperature for the AB block copolymer 57.

To understand the transition behavior, it is helpful to first establish the nature of the continuous, or flowing phase. Since poly(ethylene oxide) is crystalline (T_m ≈ 65°C) and polystyrene is amorphous (T_g ≈ 100°C) they have vastly different energies of activation for viscous flow, E_a in the temperature range of interest. Consequently, a plot of log viscosity (at zero shear rate) - vs - reciprocal absolute temperature, should give some indication of whether the ethylene oxide or styrene component dominates the flow behavior. Generally, steady state viscosity values from the Newtonian region are used so that the results are not affected by shear rate. However, styrene-ethylene oxide, in common with other block copolymers, does not show Newtonian behavior (except under specific conditions to be discussed later). For the styrene-buta-diene-styrene, SBS, system Arnold and Meier (10) have used the experimental shift factor, a_T, from time-temperature superposition of dynamic viscosity in place of steady state shear viscosity at zero shear rate. Their results show two linear plots of log a_T - vs - $1/T$ with compositions below 31% S following the line of lower slope, E_a, while compositions above 35% S follow a second line of

higher slope. Their interpretation is that below 31% S, the butadiene phase is continuous while above 35% the styrene phase is
semi-continuous and dominates the flow properties of the melt.

A similar result is found for styrene-ethylene oxide block
copolymers, except that the change in flowing phase takes place
somewhere in the composition region 20-30% S. Figure 4 for AB, and
Figure 5 for ABA block copolymers illustrate the flow behavior at
a shear rate of 0.027 sec^{-1}. Two distinct sets of curves are observed in each case. In compositions corresponding to the lower
slope ethylene oxide is the flowing phase, while at higher styrene
contents, the styrene segments dominate flow as evidenced by the
higher E_a. Because the viscosity is not Newtonian at this shear
rate no meaning can be attached to the subtle differences in E_a
values within a given set. It is, however, noteworthy that
generally the energy of activation for viscous flow observed in the
ABA series is lower (by about a factor of two) than that for the
corresponding member of the AB series. This is taken as evidence
that the ABA phase separated aggregate structures are more easily
broken down.

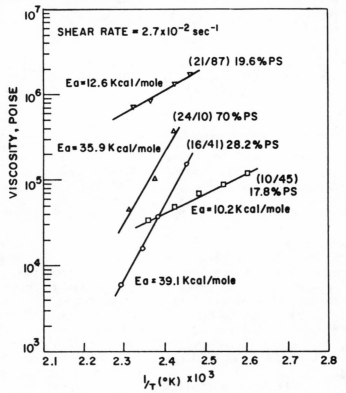

Fig. 4 - An Arrhenius plot of log viscosity for various PEO/PS
block copolymers-vs-reciprocal temperature. The viscosity data
were obtained at a shear rate of 0.027 sec^{-1}.

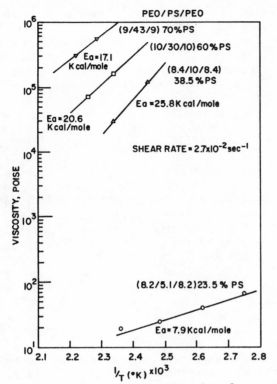

Fig 5 - An Arrhenius plot of log viscosity for various PEO/PS/PEO block copolymers-vs-reciprocal temperature. The viscosity data were obtained at a shear rate of 0.027 sec⁻¹.

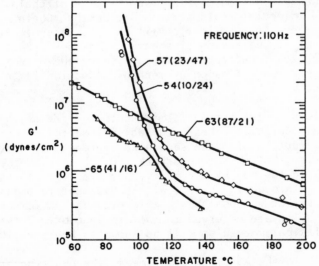

Fig. 6 - Comparison of G'(T) as functions of composition and molecular weight for several AB block copolymers of Table I.

The degree to which the styrene T_g is manifested mechanically depends principally upon the extent to which styrene forms the continuous phase, as can be seen in Fig. 6. In copolymer 63 (19.6 wt. % S) as with copolymer 61 (17.8 wt. % S) the ethylene oxide forms the continuous phase and, mechanically, the styrene transition is not observable. Tan δ, a more sensitive measurement of transition, was observed to be independent of temperature between 65° and 110°C for this composition. Since tan δ is of the order of unity in this region, the contribution of the styrene segments may be obscured. For copolymer 54 (70 wt. % S) G' decreases rapidly near 100°C, as is the case for copolymer 57 (66.7 wt. % S). At an intermediate composition, copolymer 65 (28.2 wt. % S), the styrene transition is just beginning to become noticeable. For compositions high in ethylene oxide (copolymer 61 and 63) the melt is greatly reinforced by the presence of the strongly interacting styrene phase. In the region above about 160°C, the melt elasticity appears to be most effected by the polystyrene block length, in that the G' level of copolymer 57 ($\bar{M}_n$(PS) = 47K) is higher than for copolymer 54 ($\bar{M}_n$(PS) = 24K). This is also the case for copolymer 63 ($\bar{M}_n$(PS) = 21.0K) compared with copolymer 61 ($\bar{M}_n$ = 9.73K), not shown.

C. Melt Stability

The term melt stability as used here refers to the resistance of the melt to undergo abrupt changes in modulus or viscosity with increasing temperature or shear rate. Such abrupt changes usually signify a change in the basic flow unit and may be reversible (usually delayed in time) or irreversible. Principal molecular mechanicms for changes in flow may involve slippage of chain entanglements and molecular chain scission, or both. Chain scission may occur due to autoxidation (if the presence of air is not excluded) or to the action of shearing stresses. Both types of scission have been reported for poly(ethylene oxide) (2,16,17). Polymer chain scission by mechanical action is an irreversible effect.

In what follows a distinction is made between irreversible shear induced chain scission, shear degradation, on the one hand, and reversible disruption of phase separated aggregate structures as proposed by Meier, Holden and coworkers (10,11) on the other. The disruption of aggregates may be accompanied by transfer of segments of one phase through the other, with subsequent reformation of the aggregate structure upon the release of shearing forces. Both types of behavior are observed in the ethylene oxide/styrene systems. The principal mechanism will be shown to be reversible.

In comparison to the AB block copolymers, the ABA type are less stable in the melt. Evidence for this is shown in the temperature dependence of the shear modulus (Figure 7) where in each case of similar composition the G' for the ABA lies below that for the corresponding AB. Even with the higher $\bar{M}n(PS)$ of 29.8K, ABA copolymer 31 shows substantially lower melt elasticity than co-polymer 54, $M_n(PS)$ = 24.4K.

A downward turn in G'(T), the temperature dependence of the storage modulus, observed for the ABA block copolymers at the highest measurement temperatures for each, is evidence of a structural breakdown in the melt.

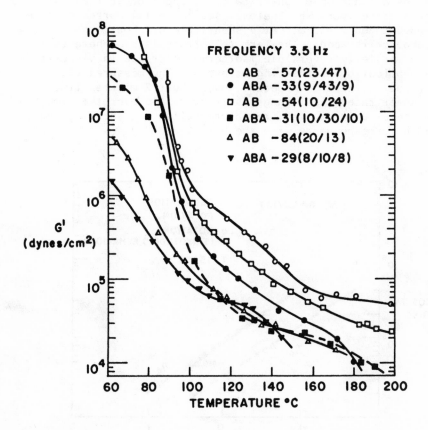

Fig. 7 - Comparison of G'(T) for AB and ABA block copolymer pairs of similar composition and molecular weight.

The G'(T) of the ABA copolymer 29 lies closer to its AB analog
than is true in the other two cases, but this is likely due to the
difference in preparation technique wherein copolymer 29 was precipitated with the low solubility parameter solvent, diethylether
(δ = 7.4). Again, the degree of intermolecular dispersion in the
melt appears to be enhanced by treatment with a liquid having a low
solubility parameter (recall sample 61C of Figure 2).

Evidence for shear degradation and structural breakdown of
the AB and ABA copolymers was obtained on the Rheo. Some typical
results for the AB block copolymer 84 are shown in Figure 8 where
the steady state viscosity, η', and the apparent viscosity, η_a, from
dynamic measurements, calculated by the method of Merz and Cox (18),
are shown as a function of shear rate, $\dot{\gamma}$, and angular frequency,
ω. η' and η_a agree very well below 1 sec^{-1}. A 1:1 correspondence
is assumed between $\dot{\gamma}$ and ω. At a shear rate of about 3 sec^{-1} thixotropic behavior is seen. The term thixotropy is used here to
indicate a partially reversible isothermal decrease in viscosity
upon the application of shearing stresses. The measured viscosity
drops by about a third of a decade, over about 2 minutes time at
constant shear rate. The change is irreversible over the time
scale of the experiment. η_a, remeasured after the shearing, again
is in close correspondence with the new η'.

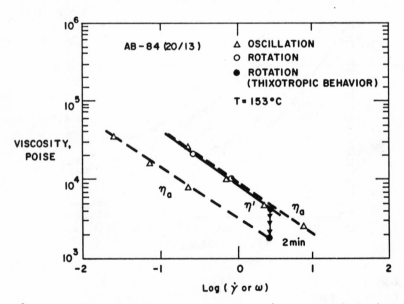

Fig. 8 - Dynamic, η_a, and steady state, η', shear viscosity behavior at 153°C for the AB block copolymer 84 as a function of shear
rate, $\dot{\gamma}$, and frequency, ω. Structural degradation by steady state
shearing is shown in the η' behavior (●) at 3 sec^{-1}.

For the ABA block copolymer 29, η' is nearly a decade below η_a at the corresponding frequency and temperature (Figure 9). Shearing at higher rates again causes a lowering of the η_a values, and upon remeasuring the steady state viscosity, Newtonian behavior begins to be observed at low shear rates.

The partial reversible character of the structural breakdown was examined in a subsequent experiment. A sample of ABA block copolymer 31, (10/30/10) 60% styrene by weight, was run at 170°C at a shear rate, $\dot{\gamma}$, of 30 sec^{-1} for one minute. The time dependence of the steady state shear viscosity, η', was then monitored at a shear rate of 8.5×10^{-3} sec^{-1}, a shear rate at which copolymer 31 is Newtonian immediately after being subjected to high shear rates. The results, Figure 10, show an increase in viscosity (measured at the low shear rate) with time, and indicate that a partial recovery of structure takes place. The original viscosity, before shearing at 30 sec^{-1} was not reached even after long times demonstrating that some irreversible shear induced polymer chain cleavage also takes place. The major portion of the shear induced viscosity decrease is, however, seen to be recoverable.

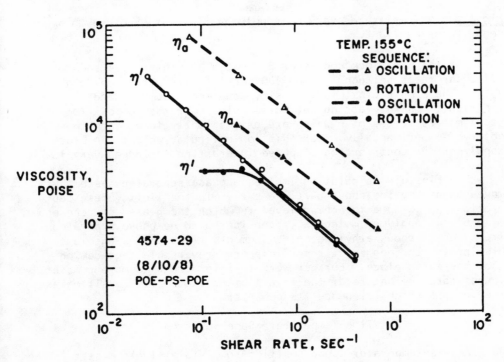

Fig. 9 - Dynamic, η_a, and steady state, η', shear viscosity behavior at 155°C for the ABA block copolymer 29 as a function of shear rate, $\dot{\gamma}$, and frequency, ω.

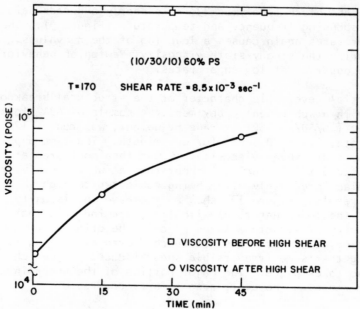

Fig. 10 - Viscosity recovery of copolymer 31 after exposure to high shear rates at 170°C.

For the ABA block copolymer 25 Newtonian behavior is seen at all shear rates between 1.0 and 100 sec^{-1} for temperatures down to 49°C. The simple Einstein viscosity equation assuming hard, spherical, non-interacting particles, gives for the composition of copolymer 25 values of η' which are only slightly lower than those observed experimentally. For this low molecular weight ABA block copolymer the contribution of phase interaction is indeed very low.

The Einstein calculation implies a phase separated system but makes no assumption about particle size. In the limit, phase separation may be on a molecular level in which the glassy styrene end of a block copolymer molecule is considered to be phase separated from the amorphous ethylene oxide segment. Thus one is unable to distinguish the degree of reversible or irreversible degradation that has taken place. Further work is in progress to more completely investigate the nature of the reversible thixotropic transition and the factors which influence its behavior.

D. Time-Temperature Superposition

A time-temperature superposition to yield a single master curve was obtained for the dynamic moduli data of copolymer 63 taken at different temperatures. Similar superposition has previously been observed for other block copolymers, such as those from styrene and butadiene (10,19), and from styrene-isoprene (20).

In the present case, the temperature dependence of the experimental shift factor, a_T, is less temperature dependent than predicted by the WLF expression (21).

CONCLUSIONS

From the results of this study, the following conclusions are drawn:

1) Melt viscosity and melt elasticity of styrene-ethylene oxide block copolymers are much larger (sometimes by orders of magnitude) than those of either homopolymer of like molecular weight or blends of these homopolymers. This is a common feature of block copolymers in general and stems from the constraints of partial phase separation.

2) In common with other block copolymers, a reversible 'transition' or change in flow behavior is observed to occur in PEO/PS and PEO/PS/PEO melts. This transition is thixotropic (shear thinning) in nature and depends upon shear rate and temperature as well as molecular parameters.

3) Thixotropy is manifested more readily in PEO/PS/PEO triblock copolymers than in PEO/PS diblocks. Lowering the molecular weight favors the onset of thixotropy at lower temperatures and shear rates.

4) In the composition range increasing from 20 to 30% styrene by weight, the continuous, or flowing phase, passes from ethylene oxide to styrene, as evidenced by an abrupt increase in the energy of activation for viscous flow.

5) When ethylene oxide forms the continuous phase, the coupling between phases is such that the glass transition of the styrene phase is not mechanically observable. This is unusual behavior for a block copolymer which is phase separated and has to our knowledge only been observed in the ethylene oxide/styrene system.

6) Time-temperature superposition has been accomplished on the styrene/ethylene oxide block copolymers using shear moduli data. In common with other block copolymers, the experimental shift factors do not obey the classical WLF correspondence which is widely observed for homopolymers.

REFERENCES

1)   I. A. Wiehe, R. G. Crystal and P. F. Erhardt, in preparation.

2)   H. Vink, Makromol. Chem., 67, 105 (1963).

3)   A. Skoulios, G. Finaz and J. Parrod, Comptes rendus, 251, 739
      (1960).

4)   A. Skoulios and G. Finaz, Comptes rendus, 252, 3467 (1961); J.
      Chim. Physique, Physico-Chim, biol., 473 (1962).

5)   E. Franta, A. Skoulios, P. Rempp and H. Benoit, Die Makromol.
      Chemie, 87, 271 (1965).

6)   D. J. Meier, Paper presented at meeting of the American
      Physical Society, Div. of High Polymer Physics, Philadelphia,
      Pa., March 1969; Block Copolymers, (J. Polymer Sci., C, 26),
      J. Moacanin, G. Holden and W. N. Tschoegl, Eds., Interscience,
      New York, 1969, p. 81.

7)   S. Krause, Paper presented at meeting of American Physical
      Society, Div. of High Polymer Physics, Philadelphia, Pa.,
      March 1969.

8)   T. Inoue, T. Soen, T. Hashimoto and H. Kwai, Paper presented
      at the IUPAC meeting, International Symposium on Macromolecular
      Chemistry, Toronto, Ont., Canada, 1968.

9)   L. Marker, Paper presented at Symposium on Block Polymers,
      American Chemical Society, New York, 1969. Polymer Preprints
      10 (2), 524 (1969).

10)   K. R. Arnold and D. J. Meier, 5th International Meeting of the
      Rheology Society, Tokyo-Kyoto, Japan, November 1968.

11)   G. Holden, E. T. Bishop and N. R. Legge, Block Copolymers, (J.
      Poly. Sci., C, 26), J. Moacanin, G. Holden and N. W. Tschoegl,
      Eds., Interscience, New York, 1969, p. 37.

12)   J. J. O'Malley, R. G. Crystal and P. F. Erhardt, This volume,
      p. 163.

13)   R. G. Crystal, P. F. Erhardt and J. J. O'Malley, This volume,
      p. 179.

14)   P. A. Small, J. Appl. Chem., 3, 71 (1953).

15)   E. Perry, J. Appl. Polymer Sci., 8, 2605 (1964).

16) W. K. Asbeck and M. K. Baxter, 'Chain Rupture of High Molecular Weight Poly(ethylene oxide) in a Uniform Shear Field', Paper, Am. Chem. Soc., Polymer Div. Meeting, Chicago, Ill., Sept. 1958.

17) R. D. Lundberg and R. W. Callard, unpublished results (1965); reported by F. W. Stone and J. J. Stratta, Encyclopedia of Polymer Science and Technology, Vol. 6, H. E. Mark and H. G. Gaylord, Eds., Interscience, New York (1967) p. 134.

18) W. P. Cox and E. H. Merz, J. Polymer Sci., 28, 619 (1958).

19) G. L. Wilkes and R. S. Stein, Paper presented at the IUPAC Meeting, International Symposium on Macromolecular Chemistry, Toronto, Ont., Canada, 1968.

20) J. F. Beecher, L. Marker, R. D. Bradford and S. L. Aggarwal, Block Copolymers (J. Polymer Sci., C, 26), J. Moacanin, G. Holden and N. W. Tschoegl, Eds., Interscience, New York, 1969, p. 117.

21) J. D. Ferry, Viscoelastic Properties of Polymers, John Wiley & Sons, Inc., New York (1961) p. 219.

ACKNOWLEDGEMENT

The authors wish to thank Stephen Strella and James M. O'Reilly for their comments throughout the course of this work. Helpful discussion with Anton Peterlin of Camille Dreyfus Laboratory is also gratefully acknowledged. Much of the experimental data was obtained through the able assistance of W. Conrad Richards.

SEGMENTED POLYURETHANES: PROPERTIES AS A FUNCTION OF SEGMENT SIZE AND DISTRIBUTION

L. L. Harrell, Jr.

Elastomer Chemicals Department

E. I. du Pont de Nemours & Co., Inc., Wilmington, Del.

Segmented polyurethanes represented by formula I are block copolymers composed of alternating soft and hard segments. They exhibit properties characteristic of crosslinked elastomers over

$$-G \underbrace{\parallel^O_{C-N} \bigcirc N}_{\text{Soft Segment}} \underbrace{[\parallel^O_{C-B-C} \parallel^O_{N} \bigcirc N]_n \parallel^O_{C}-}_{\text{Hard Segment}}$$

G = $(OCH_2CH_2CH_2CH_2)_x O-$

B = $-OCH_2CH_2CH_2CH_2O-$

I

a wide temperature range but, at higher temperatures, melt and can be processed by techniques used for plastics and fibers[1]. This behavior is explained by the fact that the polymers possess a three-dimensional network built up by intermolecular association (crystallization) of the hard segments. It was therefore of interest to study the effect upon polymer properties of changes in hard segment size, distribution, and spacing along the polymer chain.

Polymers of formula I are most conveniently prepared by the chain extension of a mixture of chloroformates with piperazine in the presence of an acid acceptor. The structures of the resulting polymers are somewhat ill-defined, however, since the average size

$$Cl-\overset{O}{\underset{\parallel}{C}}-G-\overset{O}{\underset{\parallel}{C}}-Cl + n\ Cl-\overset{O}{\underset{\parallel}{C}}-B-\overset{O}{\underset{\parallel}{C}}-Cl + (n+1)\ HN \bigcirc NH \longrightarrow I \qquad (1)$$

and size distribution of the hard segments are functions of n. A

213

further complication arises from the fact that some of the soft segments are connected by the [linkage structure] linkage which does not function as a hard segment and which serves only to alter the size distribution of the soft segments in the starting bischloroformate.

In order to separate the effects of the various structural features, it was decided to synthesize model polymers by special techniques which would permit independent control of each feature. The observed properties were then correlated with the features known to be present.

RESULTS AND DISCUSSIONS

Effect of Hard Segment Size Upon Thermal Stability of Polymer Network.

For this study a series of polymers with monodisperse hard segment size distribution was desired. To satisfy the requirement that all hard segments within a given polymer be of the same size, it was decided to first synthesize the hard segments in a stepwise manner and then to introduce them into the polymer via a condensation reaction which would preclude the possibility of any changes occurring in segment size during the chain extension step. The model polymers (I-a,b,c,d) were thus prepared by reacting amine-terminated hard segments (III-a,b,c,d) with polytetramethylene ether glycol bischloroformate (II) in the presence of an acid acceptor (eq. 2).

$$ClC\text{-}G\text{-}CCl + H\text{-}N \bigcirc N\text{-}[\text{-}C\text{-}B\text{-}C\text{-}N \bigcirc N\text{-}]_n\text{-}H \xrightarrow{Na_2Co_3} \qquad (2)$$

$$\text{II}$$

III-a, n = 1
III-b, n = 2
III-c, n = 3
III-d, n = 4

$$\text{-}G\text{-}C\text{-}N \bigcirc N\text{-}[\text{-}C\text{-}B\text{-}C\text{-}N \bigcirc N\text{-}]_n\text{-}C\text{-}$$

I-a, n = 1   , $G = (O\text{-}CH_2CH_2CH_2CH_2)_{13.7} O\text{-}$
I-b, n = 2
I-c, n = 3
I-d, n = 4

This reaction is fast, relatively free of side reactions, and produces high molecular weight polymers provided the reactants are bifunctional and the stoichiometry is properly balanced[2].

Very high quality II is easily available from commercial poly-tetramethylene ether glycol (PTMEG) by treatment of the latter with excess phosgene. III-a,b,c,d have not been prepared previously. The successful synthesis of these compounds in a high state of purity and with a high degree of bifunctionality therefore consti-tuted the major obstacle to be overcome in the synthesis of the desired polymers.

III-a,b,c,d were synthesized by the routes indicated in Figure 1. Details will be published elsewhere[3]. These compounds had elemental analyses quite close to the calculated values and amine titers between 98 and 100% of the theoretical value. They were also essentially bifunctional as indicated by the fact that the polymers prepared from them (eq. 2) had high inherent viscosities.

Properties of the model polymers are shown in Table I.

Thermal stability of the polymer networks in I-a,b,c,d were investigated by differential thermal analysis (DTA) and differential scanning calorimetry (DSC). DSC scans are shown in Figure 2. The hard segments exhibit sharp, endothermic, fusion peaks. Segment melting point increases with the number of repeat units in the seg-ment and approaches asymmtotically the melting point of the homo-polymer of the same composition as the hard segment repeat unit.

Flory[4] has derived equation (3) to describe the melting points of homopolymers of different degrees of polymerization where T_m is

FIGURE 1

SYNTHESIS OF AMINE-TERMINATED HARD SEGMENTS

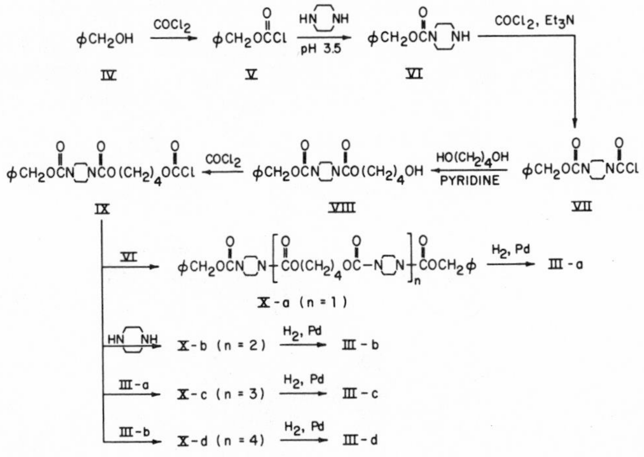

TABLE I

PROPERTIES OF SEGMENTED POLYURETHANES WITH MONODISPERSE HARD SEGMENT DISTRIBUTIONS

POLYMER	% N FOUND	% N CALC'D	INHERENT VISCOSITY[1]
I - a	4.09	4.09	2.52
I - b	5.26	5.18	2.61
I - c	6.07	6.08	2.59
I - d	6.80	6.75	2.13

[1] m - CRESOL, 0.1%, 30°C

FIGURE 2

DIFFERENTIAL CALORIMETRIC SCANS OF POLYMERS WITH MONODISPERSE HARD SEGMENT SIZES

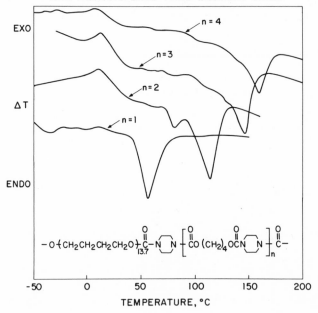

the melting point, R the gas constant, n the number of repeat units, $\bar{H}_u$ the average heat of fusion per repeat unit, and $T_m^{\,o}$

$$\frac{1}{T_m} = \frac{2\,R}{n\,\bar{H}_u} + \frac{1}{T_m}\,o \qquad\qquad (3)$$

the melting point of the infinite polymer. This equation is based on the assumption that chain ends function as diluents and thus serve to lower the melting point depending upon their concentration. The same arguments could also be applied to the hard segments in the segmented polyurethanes if the hard and soft segments were not miscible. It has not been firmly established that this condition exists although it has been observed that the melting point of a given hard segment does not change appreciably as the hard segment concentration changes. An excellent correlation of melting point versus hard segment size is obtained, however, by plotting reciprocal absolute melting temperature versus the reciprocal of the number of repeat units in the hard segment as shown in **Figure** 3.

Heats of fusion of the model polymers (III-a,b,c,d) were determined from the areas of the fusion peaks in the DSC scans. These values could then be related to the hard segment since composition

FIGURE 3

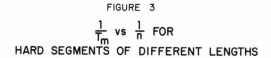

HARD SEGMENTS OF DIFFERENT LENGTHS

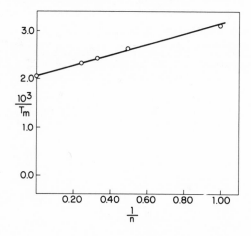

of each polymer was accurately known. The experimental heats of
fusion (Table II) are quite similar. This indicates that fractional
crystallinity of the hard segments varies little with size.

Melting Behavior of Mixtures of Different Hard Segments

Since the segmented polyurethanes (I) made under practical
conditions (eq. 1) contain hard segments of many different lengths[5],
it was of interest to study the interaction of hard segments of different size and to determine whether or not co-crystallization occurs.

Various combinations of the model polymers (I-a,b,c,d) containing equimolar quantities of the different size segments were examined.
The mixtures were prepared by dissolving the polymers in chloroform
and casting films.

Some combinations produced clear, transparent films similar
to those obtained from the individual polymers while others produced white, opaque ones.

DSC traces of the opaque films are shown in Figure 4. Fusion
peaks of the individual hard segments are present in these mixtures,
but no new ones are evident.

Traces of the clear films all showed new fusion peaks as illustrated in Figure 5. A summary of fusion peaks for the various

TABLE II

HEATS OF FUSION OF HARD
SEGMENTS OF DIFFERENT LENGTHS

NO. OF REPEAT UNITS IN HARD SEGMENT	HARD SEGMENT WT. %[1]	ΔH_f, cal/g	
		POLYMER	HARD SEGMENT
1	27.2	2.7	9.9
2	37.6	3.2	8.5
3	46.0	4.4	9.6
4	51.6	5.0	9.7
HOMOPOLYMER	100	11.4	11.4

[1]BASED ON THE UNIT

$$-\overset{O}{\overset{\|}{C}}N-N\left[-\overset{O}{\overset{\|}{C}}O(CH_2)_4O\overset{O}{\overset{\|}{C}}N-N\right]_n\overset{O}{\overset{\|}{C}}-$$

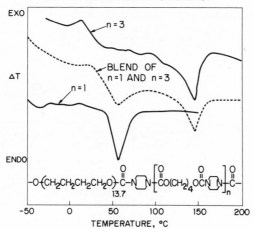

FIGURE 4
DSC SCANS OF INDIVIDUAL POLYMERS
AND PHYSICAL BLEND (OPAQUE)

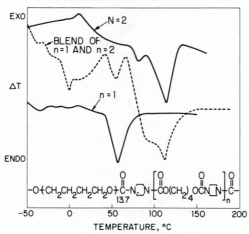

FIGURE 5

DSC SCANS OF INDIVIDUAL POLYMERS
AND PHYSICAL BLEND (TRANSPARENT)

combinations is given in Table III. All of the clear films had peaks characteristic of the two individual hard segments present and, in addition, at least two new peaks. The latter are considered to be depressed melting points indicative of cocrystallization. It is significant that the hard segment with one repeat unit cocrystallizes with a segment containing 2 units but not with segments containing 3 or 4 units. Whether or not this behavior would apply to combinations of segments containing 2,3 or 4 units with a second very long hard segment is not known.

Effect of Segment Size Distribution Upon Polymer Properties

In order to study the effects of segment size distribution, a series of polymers with the same gross composition but with independent variations in size distribution of the two different segments was synthesized as shown in Figure 6.

A sample of PTMEG was converted into a bischloroformate by treatment with phosgene. One-half of this glycol was then reacted with amine-terminated, preformed hard segment to produce a polymer with a monodisperse hard segment size distribution which was representative in this series of a narrow distribution. A polymer with the same soft segment size distribution but with broader hard segment size distribution was prepared by reacting the other half of the bischloroformate with carbobenzyloxypiperazine (VI). This reaction served the purpose of capping the glycol and preserving the molecular weight distribution of the polytetramethylene ether units present in the original glycol. After removing the carbobenzyloxy protective groups by hydrogenolysis, the amine-terminated derivative of the starting glycol was mixed with piperazine, and the mixture was chain-extended with 1,4-butanediol bischloroformate (XI). Since the two different amines are in competition for the bischloroformate in the latter reaction, the sizes of individual hard segments are statistically determined[5], and the size distribution is broad compared to that of the first polymer.

In the above reaction sequences, the size distribution of the polytetramethylene ether segments in the polymer is the same as in the starting glycol. Thus, soft segment size distribution can be conveniently varied by proper choice of the molecular weight distribution of the PTMEG. A sample of commercial PTMEG which was known to have a broad molecular weight distribution was used for preparation of model polymers containing broad soft segment size distributions. Polymers with narrower distributions were obtained by using a fractionated sample of PTMEG of the same number average molecular weight as above.

Properties of the four model polymers containing various combinations of broad and narrow distributions of the two types of segments are shown in Figure 6. Gross compositions are similar as

TABLE III

FUSION PEAKS OF POLYMER BLENDS
CONTAINING HARD SEGMENTS
OF VARIOUS SIZES

MODEL POLYMER BLEND COMPOSITION[1]	FUSION PEAKS CHARACTERISTIC OF INDIVIDUAL HARD SEGMENTS,°C				NEW FUSION PEAKS, °C			
	1	2	3	4				
1 & 2	54	112	--	--	0	90	--	--
1 & 3	54	--	146	--	-	--	--	--
1 & 4	55	--	--	160	-	--	--	--
2 & 3	--	112	144	--	-	64	102	--
2 & 4	--	112	--	160	-	63	--	147
3 & 4	--	--	146	--	-	--	107	151

[1] NUMBERS REFER TO NUMBER OF REPEAT UNITS IN INDIVIDUAL HARD SEGMENTS. BLENDS CONTAIN EQUIMOLAR QUANTITIES OF EACH HARD SEGMENT.

FIGURE 6

PREPARATION OF POLYMERS WITH
DIFFERENT SEGMENT SIZE DISTRIBUTIONS

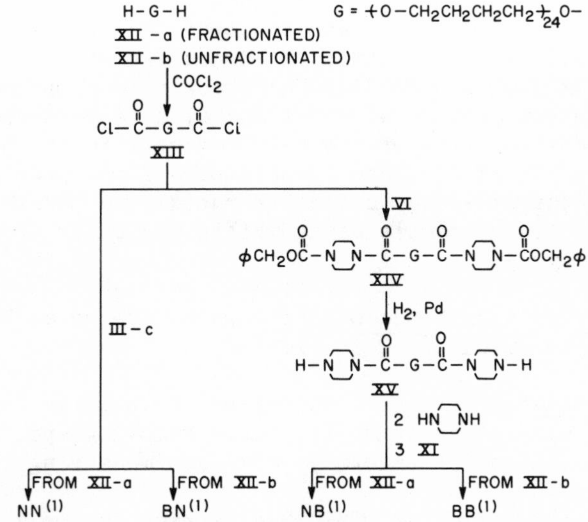

POLYMER	NN[1]	BN[1]	NB[1]	BB[1]
% N	4.32	4.28	4.41	4.26
η inh, m−CRESOL	2.62	3.89	2.03	3.78

[1] THE FIRST LETTER REFERS TO SOFT SEGMENT SIZE DISTRIBUTION; THE SECOND TO HARD SEGMENT SIZE DISTRIBUTION. N = NARROW. B = BROAD.

indicated by nitrogen content. Inherent viscosities are not the
same but are nevertheless high enough that the differences did not
affect the polymer properties investigated.

Stress-strain properties of these polymers are shown in Figure
7. It is obvious that hard segment size distribution has a profound
effect upon stress at a given strain since the narrow distribution
produces a much higher value than that of the broad distribution.

The main effect of soft segment size distribution is upon
elongation and tensile strength with the narrow distribution producing the greater values of each.

Figure 8 shows the effect of size distribution upon extension
set properties. The highest sets are produced where both segments
have narrow distributions and the lowest where both segments have
broad distributions.

EXPERIMENTAL SECTION

A. DTA and DSC Measurements were made on an apparatus similar
in design to the Du Pont Model 900 Differential Thermal Analyzer.
It was equipped with cells and the calorimetric attachment of the
Model 900. Polymer samples consisted of porous crumbs. The refer-
ence was glass for DTA measurements and air for DSC measurements.
The programmed heating rate was 12°C./min.

B. Test Specimens for stress-strain measurements consisted
of 0.25" x 5.0" x 0.020" strips died out from compression molded
slabs (3.0" x 5.0" x 0.020"). The slabs were compression molded
at a platten pressure of 40 M psi and at a temperature 25-30°C.
above the melting point of the hard segments in the polymer. The
samples were heated at these temperatures for approximately 5 minutes
after which the press was cooled as rapidly as possible with cold
water to 25°C., the sample being maintained under pressure during
the cooling cycle.

C. Stress-strain Measurements were made on an Instron machine
at a crosshead speed of 20 in./min.

D. Extension Sets were determined by pulling the specimen to
the desired elongation at 20 in./min., removing specimen from clamps.
allowing 5 minutes for recovery, and then measuring the length.

E. Fractionation of PTMEG was carried out on a belt fraction-
ator using solvent/non-solvent mixtures composed of isopropanol and
water. The original glycol ($\bar{N}_n$=1003) was separated into 10 fractions,
and the fraction with $\bar{N}_n$=1756 which represented 16.0% of the original
PTMEG was used in preparing polymers NN and NB.

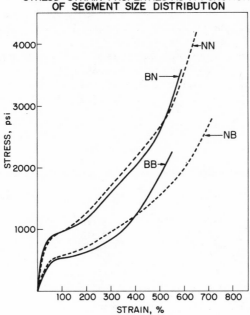

FIGURE 7

STRESS STRAIN PROPERTIES AS A FUNCTION
OF SEGMENT SIZE DISTRIBUTION

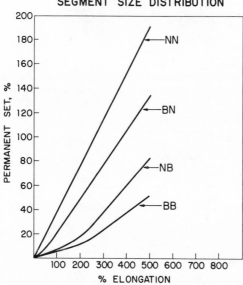

FIGURE 8

EXTENSION SET PROPERTIES
AS A FUNCTION OF
SEGMENT SIZE DISTRIBUTION

References

(1) Manfred Katz, U. S. Patent 2,929,802 (1960).
(2) P. W. Morgan, "Condensation Polymers", Interscience Publishers, Inc., New York, New York, 1965.
(3) To be published in Macromolecules.
(4) P. J. Flory, J. Chem. Phys., 15, 684 (1947).
(5) L. E. Nielsen, "Mechanical Properties of Polymers", Reinhold
Publishing Corp., New York, New York, 1962, p. 25.

MECHANICAL AND OPTICAL PROPERTIES OF BLOCK POLYMERS

II. Polyether Urethanes

G. M. Estes, D. S. Huh, and S. L. Cooper

DEPARTMENT OF CHEMICAL ENGINEERING

UNIVERSITY OF WISCONSIN — MADISON, WISCONSIN   53706

INTRODUCTION

Block polymers are a novel class of macromolecules with properties distinct from their parent homopolymers or a random copolymer of equivalent composition. Typically, an elastomeric block
polymer is built of alternate segments of a glassy or rigid polymer
and a rubbery polymer resulting in a material which is more flexible
than the pure rigid polymer, but possessing a higher modulus than
the pure rubbery segment. The intelligent design of block polymers
has opened an additional opportunity to "tailormake" molecules for
a variety of applications.

The unique and interesting properties of block polymers arise
only partly as a result of their intra-chain block structure. More
important is the fact that, with few exceptions, thermodynamic intermolecular interaction between chemically dissimilar blocks dis-
allow the formation of a true molecular solution. Rather, the
blocks prefer to be with their own kind and cluster into separate
domains. Electron microscopic studies have shown that such two-
phase structures exist, and that the size and shape of the domains
are dependent upon the composition of the block polymer and upon
the molecular weight of the constituent blocks (1-5,34). Only those
block polymers in which the two blocks are chemically very similar,
e.g., styrene-α-methyl styrene block polymers, exhibit a one-phase
structure (6,34).

The mechanical and optical properties of several block polymer
systems have proven amenable to interpretation in terms of this
observed complex microstructure. These may be broken down into
two broad classes: first, the hydrocarbon block polymers such as

styrene-butadiene-styrene and styrene-isoprene-styrene tri-block
polymers*, which are synthesized via anionic polymerization; and
second, block polymers which may be prepared by the reactions between industrially available prepolymers. Included in this class are
the segmented polyether and polyester based urethane elastomers and
block polymers prepared from the linkage of polyester or polyether
prepolymers with common glassy polymers, e.g., polystyrene or polymethylmethacrylate (7). While the first article in this series dealt
with a family of segmented polyester-based urethane elastomers (8),
in this work we are principally concerned with the effect of clustering and domain formation on the optical and mechanical properties
of polyether-based urethane elastomers.

EXPERIMENTAL

The polyether-urethane elastomers used in this study were kindly
supplied by Dr. E. A. Collins of B. F. Goodrich Chemical Company.
The soft polyether segments consisted of polytetramethylene oxide,
while the hard (aromatic urethane linkage) segments were in each case
formed using 1-4 diphenylmethane diisocyanate (MDI). Three compositions have been employed — 23.7, 31.2, and 37.7 weight % MDI. Samples
were prepared by dissolving the polymer in tetrahydrofuran and casting thin films onto a clean, mercury surface from a 5% solution.
The modulus-temperature data were obtained using the Clash-Berg (9)
and Gehman (10) torsional modulus techniques. The method employed
to gather stress-strain-birefringence data has been described
previously (8).

RESULTS AND DISCUSSION

The mechanical behavior of block polymers is characteristic of
multiphase materials, including polyblends and graft copolymers (34).
Their modulus-temperature curves exhibit two softening transition
regions which may be closely associated with the hard and the soft
segments. Between the two transitions, a region of enhanced rubbery
modulus is observed, as demonstrated by the behavior of the segmented polyether-urethanes in Figure 1. The level of modulus enhancement in the rubbery plateau region (0 to 130°C) increases with
the concentration of higher modulus block — in this case the aggregation of aromatic urethane segments. Very similar behavior has been
reported by Cooper and Tobolsky (8,11-13) for polyester-urethanes
and SBS block polymers, and by Tobolsky and Rembaum (7) for several
other block polymers.

---

*To simplify later discussions, S will be used to denote a poly-
 styrene block, B a polybutadiene block, and I a polyisoprene block.

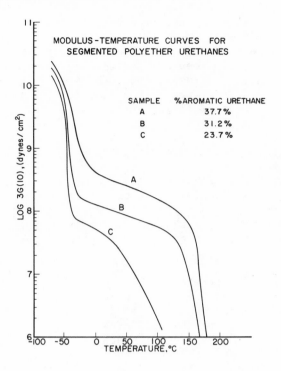

Figure 1   Modulus-Temperature Response of Polyether Urethanes.

It has been proposed that the enhanced properties are due to the hard domains functioning as virtual or pseudo-crosslinks or as filler particles which serve to reinforce the rubbery matrix (12-15). Another explanation proposed by Holden, et al. (16) is that the higher modulus domains themselves do not directly lead to the enhancement but rather act as "tie-down" points which prevent disentanglement of the rubbery matrix. Although it is fairly certain that no crystallinity is present in the hydrocarbon block copolymers so that reinforcement is due to glassy, amorphous aggregations of the hard segments, the behavior of the polyether-urethanes (17,18) and polyester-urethanes (8,11-13,17-20) has been interpreted in terms of reinforcement by micro-crystalline hard domains. However, regardless of whether crystallinity is present, it is evident that the hard domains are directly responsible for the observed increase in modulus values over those observed for unfilled, conventional vulcanizates.

The effect of clustering upon the tensile properties and the associated birefringence response of block polymers is not as readily apparent as in the case of modulus-temperature behavior. However, these properties have also been successfully interpreted in terms of domain structures. The stress-strain behavior of polyether-urethanes is displayed in Figure 2 as a function of strain history, in Figure 3 as a function of temperature, and in Figure 4 as a function of composition.

In all instances, the stress-strain curves do not obey the stress-strain relations for an ideal rubber. The cyclic stress-strain tests portrayed in Figure 2 give rise to significant hysteresis due to stress-softening. This phenomenon is common in graft copolymers (26) and filled homopolymers (24,25) as well as in other types of block polymers (2,19,27,30,34), and is characterized by a lower stress upon repeated straining up to strain levels previously reached.

The high moduli and strength exhibited by block polymers and specifically the polyether-urethanes is interpreted in terms of reinforcement by aggregations of the hard segment performing as effective crosslinks. During the initial application of stress at a given strain level, the hard domains are deformed and perhaps disrupted. Upon subsequent testing to the same strain level, the crosslinking "efficiency" of the hard domains is decreased due to the disruption and ordering, and lower moduli and stress levels result (2,19,27-29). The fact that no substantial decrease in stress-level occurs at extensions above those previously reached due to stress-softening may be easily seen by comparing the "envelope" of curves in Figure 2 with the curve at 25°C in Figure 3. In the latter case, the sample was not prestrained.

The stress-strain data of Figure 3 show that as the temperature is increased, at a given strain level, the stress on the block polyether-urethane decreases. This is in direct contrast to the behavior of an ideal rubber for which the stress is proportional to absolute temperature. Similar behavior has been reported by Fischer and Henderson (28) and Smith and Dickie (22) in SBS and SIS block polymers, by Estes, et al. (8) in polyester-urethanes, and by Riches and Haward (31) in polyester-polyether block polymers. The negative temperature coefficient of the stress arises from the fact that as the temperature is increased the hard, reinforcing phase undergoes a viscoelastic softening. Hence, the effectiveness of the higher modulus phase as a crosslink or filler particle is decreased and lower stresses result.

The modulus-temperature data presented earlier were obtained using a low strain, torsional modulus technique, and the modulus of the polyether-urethanes was seen to increase with increasing hard segment content. Figure 4 displays the corresponding higher strain

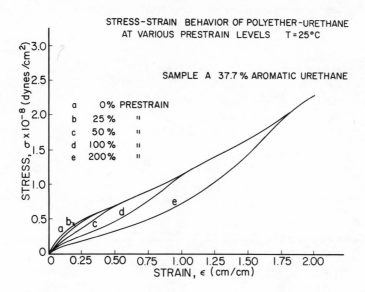

Figure 2   Stress–Strain Response of a Polyether Urethane as a
            Function of Strain History at 25°C.

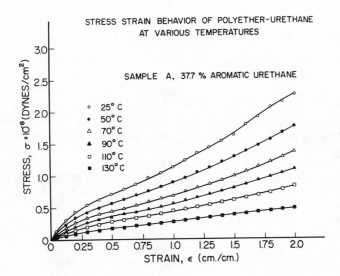

Figure 3   Stress–Strain Response of a Polyether Urethane as a
            Function of Temperature.

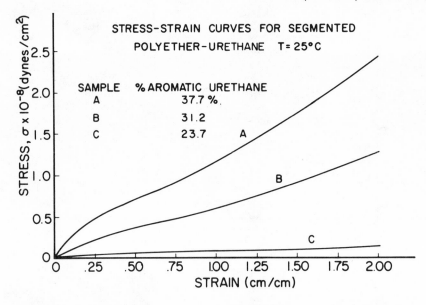

Figure 4   Stress-Strain Response of Polyether Urethanes.

data where it is seen that the modulus and stress level again increase with per cent MDI. This result is common to other block polymer systems including SBS and SIS block polymers (3,16,32), polyester
urethanes (8), and polycarbonate block copolymers (33).

The area under the stress-strain curve represents the work
which has been performed during the extension of the sample. Because stress-softening occurs in the polyether-urethane samples,
not all of the work which was put into their extension is recovered
as the samples are returned to the zero stress level. In such a
cyclic stress-strain experiment the energy which is nonrecoverable
is dissipated as heat in the stress-softening process, and is represented by the area enclosed in the hysteresis loop. The data in
Table 1 indicate that the amount of energy dissipation depends upon
the composition of the block polymer and increases with increasing
hard segment content. Further it may be observed from Table 1 that
the polyether-urethanes have lower hysteresis than the corresponding polyester-urethanes. This result is consistent with the results
of Gianatasio and Ferrari (35) and the prediction of Clough and
Schneider (17). The latter prediction was based upon the observation that a greater amount of hydrogen bonding occurs between the
urethane hydrogens and the polyester segments in polyester-urethanes
than with the polyether segments in equivalent polyether-urethane
elastomers.

TABLE 1

Nominal Aromatic Linkage Content (%)	% Energy Dissipation in Cyclic Stress-Strain Test to 100% Strain Level	
	Polyether-Urethane	Polyester-Urethane
31	32.1	46.3
38	57.0	61.4

The optical (or birefringence) data which are obtained in the tensile experiments may be interpreted similarly to the tensile data. Figure 5 shows the birefringence-stress response of the polyether-urethane at a sequence of temperatures. The nonlinearity of the curves and the fact that birefringence at equivalent stress levels increases with temperature are significant deviations from the behavior of an ideal rubber. An ideal rubber is expected to exhibit a linear plot of birefringence versus stress with a slope which decreases with increasing temperature.

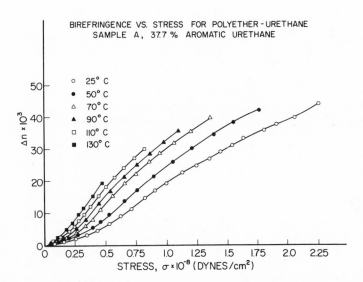

Figure 5   Birefringence-Stress Response of a Polyether Urethane as a Function of Temperature.

The birefringence-stress data from the cyclic stress-strain experiments show considerable hysteresis as presented in Figure 6. The increasing birefringence at the zero stress level accompanies a small amount of creep (or "set") after substantial sample prestrain. Similar behavior occurring as a companion to stress-softening has been reported in SIS and SBS block polymers (28,29,32), and in the polyester-urethane system (8).

Perhaps the most striking similarity in optical-mechanical properties of the various block polymer systems is the remarkable linearity in their birefringence-strain behavior up to high strain levels (greater than 200%). Figure 7 displays the data for the polyether urethane. Similar results have been reported for polyester urethanes (8,19) and SBS block polymers (21,28). While it has been demonstrated that hysteresis is present in the cyclic stress-strain and birefringence stress curves, the birefringence strain response is essentially independent of strain history as shown by the nearly equal slopes of the four curves in Figure 8. The shift of the curves relative to each other is due principally to creep of the sample during testing. The independence of strain history has been previously observed in the behavior of polyester-urethanes (8,19).

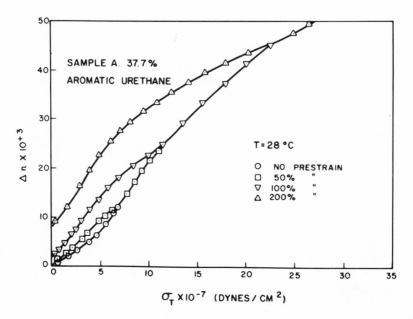

Figure 6   Birefringence-Stress Response of a Polyether Urethane
            as a Function of Prestrain.

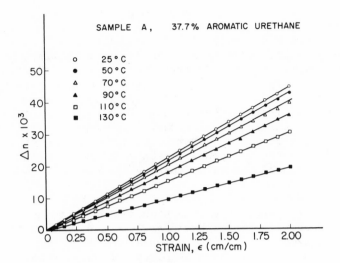

Figure 7   Birefringence-Strain Response of a Polyether Urethane as a Function of Temperature.

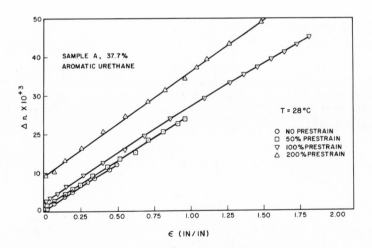

Figure 8   Birefringence-Strain Response of a Polyether Urethane as a Function of Prestrain.

CONCLUSIONS

A comparison of optical and mechanical properties of polyether-urethanes with polyester-urethanes and hydrocarbon block polymers has been made. A marked similarity is noticed in their optical and mechanical properties, reflecting a similarity of morphology in the different elastomeric block polymers. The segmented polyurethanes are not entirely similar to the hydrocarbon block polymers since calorimetric studies of both polyether- and polyester-urethanes (17, 18) as well as the birefringence work of Puett (19) and the x-ray investigations of Bonart (20) have shown that microcrystalline regions are present in these systems. The location and degree of crystallinity, of course, depends on the chemical structure, but is more common in the hard, aromatic domains.

If one adopts the domain model to describe block polymer morphology, the hypothesis that the higher modulus regions act to reinforce the rubbery networks would explain the similar mechanical properties displayed by the different systems. Further, during extension of the block polymer elastomers, the hard domains may be expected to align. In the case of the hydrocarbon block polymers these hard regions which are entirely amorphous do not significantly contribute to the birefringence (21). On the contrary, the hard domains in the polyurethane elastomers contain ordered microcrystalline regions which when oriented toward the stretching direction, yield a birefringence contribution. This additional source of birefringence could be responsible for differences in optical properties between the two types of block polymer elastomers.

ACKNOWLEDGEMENT

The authors wish to thank Dr. E. A. Collins and Mr. D. G. Frazer of the B. F. Goodrich Chemical Company for preparation of the poly-ether urethanes used in this work. The authors also wish to acknow-ledge The National Science Foundation for support of this research through Grant GK 4554.

BIBLIOGRAPHY

1. J. F. Beecher, L. Marker, R. D. Bradford and S. L. Aggarwal, Polymer Preprints, 8, 1532 (1967).

2. J. F. Beecher, L. Marker, R. D. Bradford and S. L. Aggarwal, J. Polymer Sci., C (26), 117 (1969).

3. M. Morton, J. E. McGrath and P. C. Juliano, J. Polymer Sci., C (26), 99 (1969).

4. M. Matsuo, Japan Plastics, July 1968, p. 6.

5. M. Matsuo, T. Ueno, H. Horino, S. Chujyo, and H. Asai, Polymer, 9, 425 (1968).

6. M. Baer, J. Polymer Sci., A (2), 417 (1964).

7. A. V. Tobolsky and A. Rembaum, J. Appl. Polymer Sci., 8, 307 (1964).

8. G. M. Estes, R. W. Seymour, D. S. Huh, and S. L. Cooper, to be published, Polymer Eng. and Sci. (in press).

9. ASTM Standards, American Society for Testing and Materials, Philadelphia, 1961, D 1043-61T.

10. ASTM Standards, American Society for Testing and Materials, Philadelphia, 1958, D 1053-58T.

11. S. L. Cooper and A. V. Tobolsky, Textile Res. J., 36 (9), 800 (1966).

12. S. L. Cooper and A. V. Tobolsky, J. Appl. Polymer Sci., 10, 1837 (1966).

13. S. L. Cooper and A. V. Tobolsky, J. Appl. Polymer Sci., 11, 1361 (1967).

14. R. F. Fedors, J. Polymer Sci., C (26), 189 (1969).

15. E. M. Hicks, Jr., A. J. Ultee, and J. Drougas, Science, 147, 373 (1965).

16. G. Holden, E. T. Bishop, and N. R. Legge, J. Polymer Sci., C (26), 37 (1969).

17. S. B. Clough and N. S. Schneider, J. Macromol. Sci.-Phys., B-2 (4), 553-566 (1968).

18. S. B. Clough, N. S. Schneider and A. O. King, J. Macromol. Sci.-Phys., B-2 (4), 641-648 (1968).

19. D. Puett, J. Polymer Sci., A-2 (5), 839 (1967).

20. R. Bonart, J. Macromol. Sci., B-2 (1), 115 (1968).

21. G. L. Wilkes and R. S. Stein, J. Polymer Sci., A-2 (in press).

22. T. L. Smith and R. A. Dickie, J. Polymer Sci., C (26), 163 (1969).

23. K. C. Rusch, J. Macromol. Sci., B-2 (3), 421 (1968).

24. B. B. Boonstra, Chapter 16, Reinforcement of Elastomers, Ed. G. Kraus, Interscience (1965).

25. G. Kraus, Rubber Chem. and Technology, 38 (5), 1070 (1965).

26. F. M. Merrett, J. Polymer Sci., 24, 467 (1957).

27. C. W. Childers and G. Kraus, Rubber Chem. and Technology, 40, 1183 (1967).

28. E. Fischer and J. F. Henderson, J. Polymer Sci., C (26),
     149 (1969).

29. J. F. Henderson, K. H. Grundy, and E. Fischer, J. Polymer Sci.,
     C (16), 3121 (1968).

30. S. L. Cooper, D. S. Huh, W. J. Morris, Ind. and Eng. Chem.,
     Prod. Res. and Dev., 7 (4), 248 (1968).

31. K. Riches and R. N. Haward, Polymer, 9, 103 (1968).

32. D. M. Brunwin, E. Fischer, and J. F. Henderson, J. Polymer
     Sci., C (26), 135 (1969).

33. E. P. Goldberg, J. Polymer Sci., C (4), 707 (1964).

34. G. M. Estes, A. V. Tobolsky and S. L. Cooper, J. Macromol.
     Sci., C (in press).

35. P. A. Gianatasio and R. J. Ferrari, Rubber Age, June 1966,
     p. 83.

SYNTHESIS AND PROPERTIES OF SILOXANE BLOCK POLYMERS

E.E. Bostick

General Electric Insulating Materials Department

Schenectady, New York

INTRODUCTION

Organosiloxane block polymers have not received the interest and effort which have been applied to hydrocarbon block polymers during the past several years (1). This may be attributed to the lack of synthetic technique available for the convenient preparation of linear block polysiloxanes having well-defined homopolymer segments. All of the systems disclosed in the literature involve condensation reactions between difunctional homopolymer segments to yield products having random blocks and are, at best, poorly characterized (2,3). For example:

$$
\begin{array}{cccccccc}
R & R & & R' & R' & R & R & R' & R' \\
| & | & & | & | & \text{-HCl} & | & | & | & | \\
\text{HO} -\text{SiO-} & \text{SiOH} + \text{Cl} -\text{SiO-} & \text{SiCl} & \longrightarrow -\text{Si} -\text{OSi-} & \text{OSi-} & \text{OSi-} \\
| & | & & | & | & \text{Py} & | & | & | & | \\
R & R & & R' & R' & R & R & R' & R' \\
\end{array}
$$

Side reactions such as silanol condensation in the above system may lead to cyclic products and loss of block control.

The copolymerization of organocyclosiloxanes to produce ordered blocks of interconnected polymer segments has not been disclosed in the prior literature. All acid and strong base-catalyzed polymerization techniques have been shown to cause equilibration of rings and chains and to lead to random mixtures of rings and chains (4). For example:

$$[(CH_3)_2SiO]_4 \ + \ KOH \ \longrightarrow \ HO \left(\begin{array}{c} CH_3 \\ | \\ -SiO- \\ | \\ CH_3 \end{array} \right)_4 K + [(CH_3)_2SiO]_n$$

A further complication arises in copolymerization due to the wide differences in reactivity of various organocyclosiloxanes. Studies have been reported on the reactivities of some of the cyclosiloxanes. These studies indicate the random nature of products arising from copolymerization of organocyclosiloxanes[5].

The requirements for a polymerization initiator which will allow synthesis of block polysiloxanes from organocyclosiloxanes are:

1. It must react with and open siloxane rings to cause a multi-step addition polymerization reaction.

2. The active species must be terminal to the polymer chain.

3. The active species must not undergo exchange reactions with siloxane chain segments to cause equilibrations.

4. The active species must have a lifetime sufficient to allow sequential additions of organocyclosiloxanes.

RESULTS

Synthesis

During our investigation of the polymerization of various organocyclosiloxanes, we have developed techniques, catalysts and reaction conditions which allow the unequivocal synthesis of block polymers from organocyclosiloxanes. The catalysts used for these reactions are basic lithium derivatives. For example, when hexamethylcyclotrisiloxane is reacted with the lithium salt of diphenylsilane diol in the presence of tetrahydrofuran, a multi-step addition polymerization occurs in the following manner:

1. $\quad n\left(\begin{array}{c} CH_3 \\ | \\ -SiO- \\ | \\ CH_3 \end{array}\right)_3 \; + \; \begin{array}{c} \phi \\ | \\ LiOSi-OLi \\ | \\ \phi \end{array} \xrightarrow[\text{r.t.}]{\text{THF}} \; Li\left(\begin{array}{c} CH_3 \\ | \\ -OSi- \\ | \\ CH_3 \end{array}\right)_{\frac{3n}{2}} \begin{array}{c} \phi \\ | \\ OSi-O \\ | \\ \phi \end{array} \left(\begin{array}{c} CH_3 \\ | \\ -SiO- \\ | \\ CH_3 \end{array}\right)_{\frac{3n}{2}} Li$

$n \geq 1$

Likewise, when hexamethylcyclotrisiloxane is reacted with a monofunctional lithium triorganosilanolate, the reaction proceeds:

2. $\quad n\left(\begin{array}{c} CH_3 \\ | \\ -SiO- \\ | \\ CH_3 \end{array}\right)_3 \; + \; \begin{array}{c} \phi \\ | \\ LiOSi-\phi \\ | \\ CH_3 \end{array} \xrightarrow[\text{r.t.}]{\text{THF}} \; \begin{array}{c} \phi \\ | \\ \phi-Si-O \\ | \\ CH_3 \end{array} \left(\begin{array}{c} CH_3 \\ | \\ -SiO- \\ | \\ CH_3 \end{array}\right)_{3n} Li$

$n \geq 1$

Further, when another cyclosiloxane is added to this reaction system, polymerization continues to produce block polyorgano-siloxanes of the ABBA or AB type. For example, when hexaphenyl-cyclotrisiloxane is added, the reaction may be represented as follows:

3. $\quad Li\left(\begin{array}{c} CH_3 \\ | \\ -OSi \\ | \\ CH_3 \end{array}\right)_{\frac{3n}{2}} \begin{array}{c} \phi \\ | \\ O-Si-O \\ | \\ \phi \end{array} \left(\begin{array}{c} CH_3 \\ | \\ -SiO- \\ | \\ CH_3 \end{array}\right)_{\frac{3n}{2}} Li \; + \; m\left(\begin{array}{c} \phi \\ | \\ -Si-O \\ | \\ \phi \end{array}\right)_{3m} \xrightarrow[\text{or anisole}]{\text{THF}/C_6H_6}$

$Li\left(\begin{array}{c} \phi \\ | \\ -OSi- \\ | \\ \phi \end{array}\right)_{\frac{3m}{2}} \left(\begin{array}{c} CH_3 \\ | \\ O-Si- \\ | \\ CH_3 \end{array}\right)_{\frac{3n}{2}} \begin{array}{c} \phi \\ | \\ O-Si-O \\ | \\ \phi \end{array} \left(\begin{array}{c} CH_3 \\ | \\ -Si-O \\ | \\ CH_3 \end{array}\right)_{\frac{3n}{2}} \left(\begin{array}{c} \phi \\ | \\ -Si-O- \\ | \\ \phi \end{array}\right)_{\frac{3m}{2}} Li$

The lithium derivatives that are effective as catalysts for this type of sequential block polymerization are fairly general. In addition to the difunctional and monofunctional silanolates mentioned above, other monofunctional lithium triorganosilanolates, alkyllithium, aryllithium, lithium hydroxide, lithium alkoxide, lithium hydride, and lithium aluminum hydride have been used as initiators for the initial reactive block segment.

The addition of electron-donating aprotic solvents promotes the reaction without inducing any loss of specificity or accompanying randomization. This effect is completely anomalous to previous reports that lithium ion exhibits behavior similar to sodium or potassium ion in the presence of solvents such as tetrahydrofuran (6,7). Other additives such as 1,2-dimethoxy-ethane, diethyleneglycol-dimethylether, anisole, triethylamine, and p-dioxane promote the reaction in the same manner.

Reactions conducted wherein both diorganocyclosiloxanes were added simultaneously have failed to yield copolymers which analyze or behave as block copolymers. This agrees with literature reports of reactivity of organocyclosiloxanes during copolymerization (5). Likewise, sodium or potassium derivatives failed to provide a convenient block copolymerization initiator.

The difunctional organocyclosiloxanes which have been used in the preparation of block polysiloxanes include hexamethyl-cyclotrisiloxane, hexaphenylcyclotrisiloxane, both cis- and trans-2,4,6-trimethyl-2,4,6-triphenylcyclotrisiloxane, cis-2,4-dimethyl-2,4,6,6-tetraphenylcyclotrisiloxane, 2,2-dimethyl-4,4,6,6-tetraphenylcyclotrisiloxane, and tris-cyclopentamethylene-cyclotrisiloxane. The primary types of block polysiloxanes described in this report are as follows:

$$\underline{AB}, \ \underline{ABBA}, \ \underline{ACBBCA}, \ \underline{ABCCBA},$$

$$\text{where} \quad A = [\phi_2 SiO]_x$$

$$B = [(CH_3)_2 SiO]_x$$

$$C = [\phi CH_3 SiO]_x$$

Characterization and Properties

Random copolymers and terpolymers of diorganosiloxanes have been investigated and reported in the literature for some time. In general, the properties of these materials have not been described to be unusual or different from the properties expected for a random copolymer (8). On the contrary, the block polymers described in this report exhibit unique and unusual properties. On first inspection they are tough, white, opaque elastomers which are primarily insoluble in common organic solvents at room temperature. They are soluble in such solvents as diphenylether at elevated temperatures. This resistance to dissolution is due to the crystallinity of the polydiphenylsiloxane segments. Table I lists some crystalline melting points of ABBA diphenyl/dimethyl block copolymers which have various composite ratios and diphenyl-siloxane segment length.

TABLE I

Diphenyl-Dimethyl Block Copolysiloxanes

Composition Mole %		Design Segment Length x10^3				Tm, °C	
		Dimethyl		Diphenyl			
D_3^a	P_6^b	M_n	DP_n	M_n	DP_n	ϕ_2SiO	$(CH_3)_2SiO$
73	27	150	2	75	.375	237	-42
80	20	150	2	50	.250	236	-49
84.25	15.75	150	2	37.5	.185	228	-49
91.5	8.5	150	2	18.8	.100	(210) 228	-47
96	4	150	2	8.5	.042	cooling*	-46

D_3 = hexamethylcyclotrisiloxane

P_6 = hexaphenylcyclotrisiloxane

*Crystallization was induced by cooling in DSC.

Nuclear Magnetic Resonance Spectroscopy of Block Copolydiorganosiloxanes

During the course of this work and related investigations of the chemistry of organocyclosiloxanes, high resolution nuclear magnetic resonance spectroscopy has been found to be useful in determining microstructure of both cyclic and linear polydiorgano-siloxanes (9). The established observation that pmr spectra of methyl protons in polymethylorganosiloxanes reflect the number and types of environments of methylorganosiloxy groups present will be utilized here in the illustration of structural differences between block and random copolymers. The difunctional units compared under block and random conditions were derived from the following cyclics:

1. hexamethylcyclotrisiloxane plus hexaphenylcyclotrisiloxane (Figs. 1 and 2).

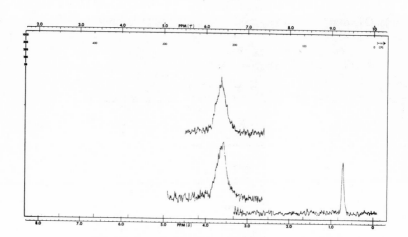

Figure 1.   The pmr spectra of methyl
protons of a dimethyl/diphenyl <u>block</u>
copolymer.

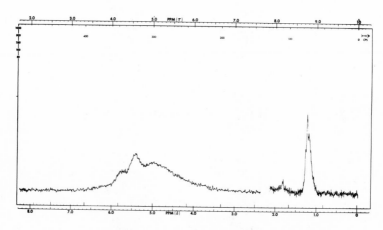

Figure 2.   The pmr spectra of methyl
protons of a dimethyl/diphenyl/random
copolymer.

2. <u>cis</u>-2,4,6-trimethyl-2,4,6-triphenylcyclotrisiloxane
    plus hexaphenylcyclotrisiloxane (Figs. 3 and 4).

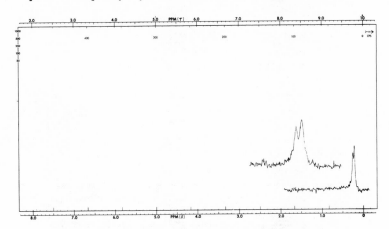

Figure 3.   The pmr spectra of methyl protons
of a <u>block</u> copolymer derived from <u>cis</u>-2,4,6-
tri-phenylcyclotrisiloxane and hexaphenylcyclotrisiloxane.

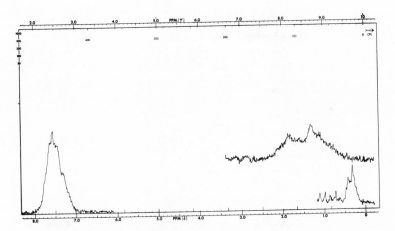

Figure 4.   The pmr spectra of the methyl protons
of a <u>random</u> copolymer derived from <u>cis</u>-2,4,6-
triphenylcyclotrisiloxane.

3. trans-2,4,6-trimethyl-2,4,6-triphenylcyclotrisiloxane
    plus hexaphenylcyclotrisiloxane (Fig. 5).

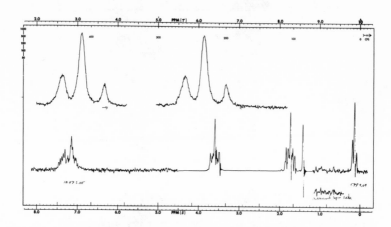

Figure 5. The pmr spectra of methyl protons of
soluble fraction of poly(methylphenylsiloxane-
block-diphenylsiloxane).

4. trans-2,4,6-trimethyl-2,4,6-triphenylcyclotrisiloxane
    plus hexamethylcyclotrisiloxane (Figs. 6 and 7).

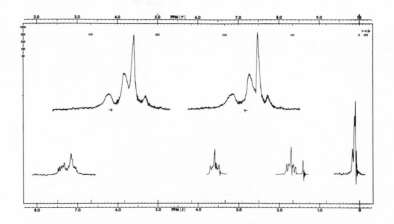

Figure 6. The pmr spectra of methyl protons of
poly(dimethylsiloxane-block-methylphenylsiloxane).

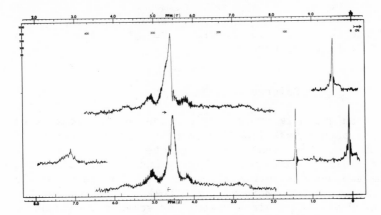

Figure 7. The pmr spectra of methyl protons
of poly(dimethylsiloxane-CO-methylphenylsiloxane)
(random from trans-trimer + D₃).

Block copolymers included in the measurements contained pre-
designed segments which were long enough to minimize end group
effects. However, the conditions of synthesis were such that if
any equilibration or randomization were significant, the pmr
spectra should reflect these reactions. For convenience, the
conclusions drawn from the spectra are listed in Table II.

TABLE II

Comparison of pmr Spectra of Block and
Random Copolyorganosiloxanes

Case	Figure	Spectra	Conclusions
1	I	Singlet	Block
1	II	Multiplet	Random
2	III	Doublet[a]	Block
2	IV	Multiplet	Random
3	V	Triplet[b]	Block
4	VI	Triplet[c] w/ singlet	Block
4	VII	Multiplet	Random

a) Agrees with cis-2,4,6-trimethyl-2,4,6-triphenylcyclotrisiloxane
b) Agrees with trans-2,4,6-trimethyl-2,4,6-triphenylcyclotri-
   siloxane
c) Agrees with integrity of polymethylphenylsiloxane segments

The conclusions drawn from pmr spectra along with other
physical and chemical evidence presented are considered adequate
for proof of block structure.

Solution and Bulk Properties

Solution and bulk properties provide an excellent demon-
stration of both crystalline and amorphous block segments in
polymer chains. The crystalline segments contribute to solvent
resistance, high temperature stability and tensile strength,
while the amorphous segments allow low temperature flexibility
and/or extensibility. Moreover, the bulk properties may be
altered somewhat by previous thermal or solvent treatment.

The poly(dimethylsiloxane-block-diphenylsiloxane) described
in this report may be obtained as crystalline elastomers from
the reaction flask. The dissolution of the polymer in hot diphenylether (190-230°C) followed by quenching and precipitation in cold
methanol has been found to transform the elastomer into a high
modulus material. A comparison of the properties of crosslinked
specimens revealed that the tensile strength was increased, the
flexibility was decreased, and the resistance to swell by common
aromatic solvents was increased.

This behavior may not be adequately explained by an increase
or change in crystalline content. Results from measurements on the
precipitated polymer by differential scanning calorimetry (DSC) do
not indicate any significant increase in crystallinity, but do show
some lowering of melting point temperatures for the diphenyl-
segments until thermally cycled.

The behavior may be explained by consideration of the effect
of solvent or environment on segment expansion. First, the
flexibility of the polymer is believed to be due to the dimethylsiloxane segments which were formed in the presence of a good
solvent and are extended to their most probable length. On the
other hand, the diphenylsiloxane segments are formed in poorly
solvating media and crystallize as soon as a reasonable segment
length is attained. Thus, the crystallites form from highly
constricted or coiled segments. The subsequent solvent treatment
expands the polydiphenylsiloxane segments to form a continuous
phase of high modulus material in the recovered polymer.

The implications of such behavior by a polymer system are
both practically and academically important. The investigation of
effects of various solvents, temperature and copolymer composition
continues in this area.

Acknowledgement

I wish to thank Dr. J.B. Bush, Jr. for much valuable
assistance in obtaining and interpreting nmr data and for many
helpful discussions. I also extend my appreciation to
Mr. K.A. Kluge for calorimetric measurements.

References

1. R.J. Ceresa, Block and Graft Copolymers, Butterworth,
    Washington, 1962.

2. R.L. Merker, M.J. Scott, and G.G. Haberland, J. Polymer
    Sci., A-2, 31 (1964).

3. U.K. Patent Specification 994,401 (Dow-Corning Corp.),
    June 10, 1965.

4. W.T. Grubb and R.C. Osthoff, J. Am. Chem. Soc., 77, 1405
    (1955).

5. R.L. Merker and M.J. Scott, J. Polymer Sci., 43, 297
    (1960).

6. K.F. O'Driscoll and A.V. Tobolsky, J. Polymer Sci., 35,
    259 (1959).

7. A.V. Tobolsky and C.E. Rogers, J. Polymer Sci., 38, 205
    (1959).

8. O.K. Johannson, U.S. Patent 2,868,766, January 13, 1959.

9. J.B. Bush, Jr. and E.E. Bostick, to be published.

AN ELECTRON MICROSCOPICAL STUDY OF THE MORPHOLOGY OF POLYSILOXANE BLOCK COPOLYMERS

A. Keith Fritzsche* and Fraser P. Price*

General Electric Research and Development Center

Schenectady, New York   12301

The domain structure of several block copolymers is well documented. The optical behavior of solutions of polyethyleneoxide-polystyrene bloc copolymers gives evidence of liquid-crystal-like structure[1]. Studies using the electron microscope to investigate these films of block copolymers, notably polystyrene-polybutadiene (2), have revealed spherical, rodlike or lameller domains depending upon the relative proportions of the two components. Theories based upon thermodynamic considerations have been proposed to explain the various morphologies[3].

The preparation of polysiloxane block copolymers of the ABA type where A is polydiphenylsiloxane and B is polydimethylsiloxane has been described by E. E. Bostick[4] and some of their properties have been delineated. Multisequence block copolymers have also been described where the molecules of ABA have been coupled to yield larger molecules of the type ABAABAABA----ABA. The blocks in both the ABA polymers and the multisequence copolymers are moieties of vastly differing properties. The A block is polydiphenyl-siloxane, which is a highly hindered structure and is a crystalline, high melting solid. The B block is polydimethylsiloxane which has a very flexible molecule and is either an oil or a viscoelastic material. This welding of moieties of such widely varying properties promised to produce materials with unusual characteristics. Accordingly the electron microscopical study described herein was

* Present Address, Polymer Science and Engineering Program,
  University of Massachusetts, Amherst, Massachusetts 01002.

undertaken. Studies were made of precipitates of the ABA block copolymers from dilute solution and of thin cast films viewed in direct electron transmission of both the ABA block copolymers and of the multisequence block copolymers. In addition a homopolymer of the polydiphenylsiloxane was also studied. It is worth noting that whereas in the microscopical studies on polystyrene-polybutadiene block copolymers it was necessary to stain the polybutadiene blocks with osmium tetroxide to get sufficient contrast, in the present study this procedure was not necessary.

Materials

The ABA block copolymers are described in Table I.

TABLE I

Composition of ABA Block Copolymers

$$A = (\emptyset_2 SiO)_x \quad B = (Me_2 SiO)_y$$

Polymer	x	y	DP=2X+y	%($\emptyset_2$SiO)
BC 1*	500	4000	5000	20
BC 2	96	2150	2340	8
BC 3	150	2200	2500	12
BC 4	190	2170	2450	16

* Values for this polymer are approximate

The polymer BC 1 unfortunately was not as well characterized as the others but it is the one with which the largest amount of work was done. The remaining polymers constitute a series of approximately constant DP but with increasing diphenylsiloxane content.

The multisequence block copolymers which will be discussed are described in Table II.

TABLE II.

Composition of Multisequence Block Copolymers

$$x = DP \text{ of } \emptyset_2 SiO \text{ Blocks}$$

$$y = DP \text{ of } Me_2 SiO \text{ Blocks}$$

Polymer	Prepolymer			Coupled		Polymer
	x	y	$[\eta]^{(a)}$	x	y	$[\eta]^{(a)}$
MSB 1	400	2440	----	800	2440	0.85
MSB 2	300	1250	0.48	610	1250	0.58
MSB 3	350	1640	0.34	710	1640	0.63

(a) Measured in o-dichlorobenzene at 120°C.

In some of the experiments MSB 1 was diluted with a poly-dimethylsiloxane polymer of DP 2400. It is to be noted that this chain length is essentially that of the dimethyl block in MSB 1.

In addition some precipitations from dilute solution were carried out with a diphenylsiloxane homopolymer of undetermined chain length. Probably the degree of polymerization was quite low since it did dissolve in hot chlorobenzene and it is known that high molecular polydiphenylsiloxanes are extremely intractable.

Dilute Solution Experiments

Dilute solution precipitation experiments were performed only with the ABA copolymer, BC 1, and with the diphenylsiloxane homopolymer. These were dissolved in monochlorobenzene at 1% concentration, diluted and cooled to room temperature. A drop of the suspension was put on a microscope slide, the solvent evaporated and the residue shadowed and replicated by conventional techniques and examined in the electron microscope.

In Figure 1 are displayed the structures observed in the precipitate of the diphenylsiloxane homopolymer. This figure shows two types of structures, (a) lathlike structures 0.1-0.2µ wide and a few angstroms thick apparently have some rigidity because firstly they are quite straight over significant distances (1-2µ) and secondly where they cross they show little tendency to bend and (b) square terraced structures. The interior details of one of these terraced structures is shown in Fig. 2. It is seen that the lathlike structure is preserved here. In other specimens of this precipitate square spiral ramps could be seen.

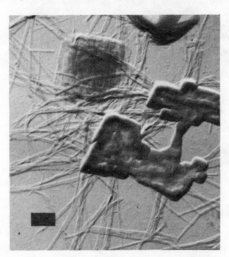

Fig. 1.   Crystals of polydiphenylsiloxane homopolymer precipitated
           from chlorobenzene solution. Pt shadowed. Marker - 1µ.

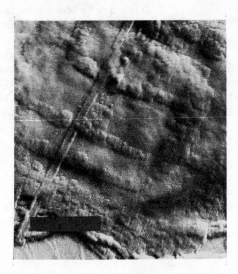

Fig. 2. Multilayer crystal of polydiphenylsiloxane homopolymer
         precipitated from chlorobenzene solution. Pt shadowed.
         Marker – 1μ.

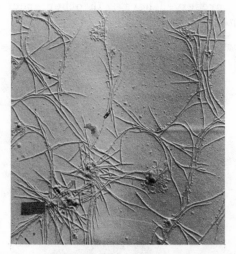

Fig. 3. Threadlike crystals of polymer BC 1 precipitated from
         chlorobenzene solution. Pt shadowed. Marker – 1μ.

The precipitate of the block copolymer BC 1 is shown in Fig. 3. Here the threads are much smaller in cross sections and seem to be more flaccid exhibiting a continuously curving structure. The fibers are about equidimensional in cross section (about 0.05μ) and are several microns long. In the precipitate of BC 1 occasionally terraced lamellar structures are found. However they are rare.

Several experiments were made where the temperature of the solution was allowed to fall from $150^{\circ}C$ to some temperature above room temperature and the precipitate filtered off at this elevated temperature. Thus some information could be developed about the effect of temperature upon the morphology. Here it was found that with BC 1 polymer; fibers were formed predominently when the temperature never dropped below $75^{\circ}C$ while lamellae predominated when the minimum temperature was about $50^{\circ}C$.

Thin Film Experiments

Films of the polymer described in Tables I and II were prepared by casting one per cent solutions in hot monochlorobenzene on nitrocellulose treated slides. These slides were maintained at $150^{\circ}C$ on an inclined hot stage in a CO_2 atmosphere. A few drops of the solution were placed on the top of the inclined slide and allowed to run down. This produced films thin enough to view in the electron microscope by direct transmission. Some films were annealed in an inert atmosphere at temperatures ranging up to $500^{\circ}C$. Only above $300^{\circ}C$ annealing temperature was a change in morphology observed.

ABA Block Copolymers. The ABA polymer BC 1 was studied most extensively. The AS cast film displayed in Fig. 4 shows numerous dark regions that are approximately equidimensional varying in size from 600 - 1300Å and essentially the same distance apart. It is clear that these dark regions are the discontinuous phase. Subsequent experiments concerned with the effect of the size and concentration of the diphenylsiloxane blocks, plus the high optical contrast, lead us to identify the dark regions as the crystallized microphase of polydiphenylsiloxane. The boundaries of the dark regions in the as-cast films are quite diffuse. Figure 5 shows the effect of annealing above $300^{\circ}C$. It is seen that the boundaries become much sharper. The contrast increases and the average dimension decreases by about a factor of two into the 300 - 650Å range. These observations are consistent with the idea that the rapid casting process produces non-equilibrium structures which can be made to relax upon annealing.

The optical contrast in these films is surprisingly high, probably as a result of Bragg scattering out of the objective apeture of the microscope. Shadowed surface replicas of these films allowed us to identify the dark regions as thin spots in the film.

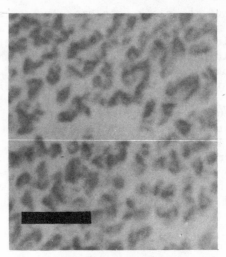

Fig. 4.   Film of BC 1 cast from chlorobenzene solution.
          Marker - 0.5μ.

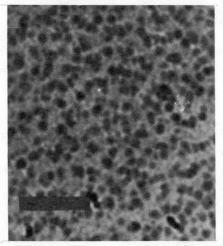

Fig. 5.   Same film as in Fig. 4 after heating for 45 minutes at
          500°C. Marker - 0.5μ.

Figures 6, 7 and 8 show micrographs of films prepared from
BC 2, BC 3 and BC 4 respectively. They indicate that as the
diphenylsiloxane block length is increased the size of the crystalline regions also increased. In these films, annealing above
300°C produced more numerous and better defined crystalline regions.
In these three polymers as opposed to BC 1, the annealed crystalline
regions appeared to be rodlike, rather than equidimensional. The
polymer BC 3 was studied most intensively. Figure 9 shows at a
higher magnification the structures of Fig. 7. Here the rods appear to

be at a quite constant width of about 400Å with lengths of 1000 -
3000Å. Furthermore the rods tend to lie in groups of three to
eight with their long axes parallel. This type of clustering of
rods seems to be typical of films of the multisequence block copolymers discussed below.

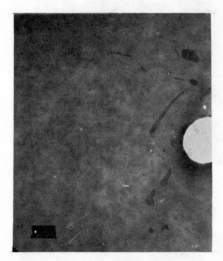

Fig. 6.   Film of BC 2 cast from chlorobenzene solution.
          Marker - 1μ.

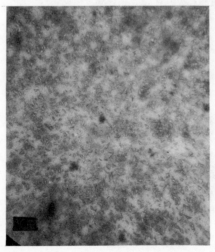

Fig. 7.   Film of BC 3 cast from chlorobenzene solution.
          Marker - 1μ.

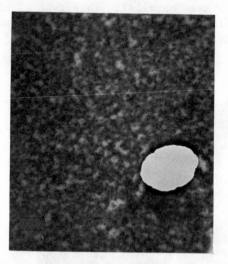

Fig. 8.   Film of BC 4 cast from chlorobenzene solution.
          Marker – 1μ.

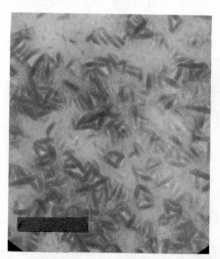

Fig. 9.   Film of BC 3 cast from chlorobenzene solution.
          Marker – 0.5μ.

Multisequence Block Copolymers. Films of the polymers described in Table II were prepared, annealed and examined by the techniques described above. Films were cast both from monochlorobenzene at 150°C and from diphenylether at 260°C. This latter process allows production of a film closer to its equilibrium condition.

Figures 10, 11 and 12 show electronmicrographs of films of MSB 1, MSB 2 and MSB 3 respectively. These polymers all have about the same diphenyl block length but vary in the viscosity of the coupled polymer and in the length of the dimethylsiloxane block. Figures 10 and 11 show the parallel rod arrays described above. For films as cast from monochlorobenzene the similarity of the ABA copolymers to the multisequence block copolymer decreases as the total chain length of the multisequence polymer increases. For example Polymer MSB 3 in Fig. 12 essentially shows no phase separation. Annealing however tends to restore the similarities.

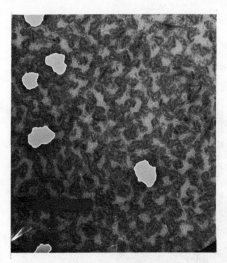

Fig. 10. Film of MSB 1 cast from chlorobenzene solution. Marker - 0.5μ.

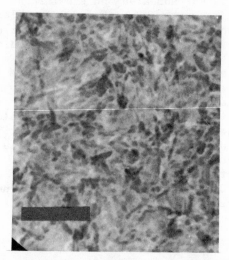

Fig. 11.   Film of MSB 2 cast from chlorobenzene solution.
Marker - 0.5μ.

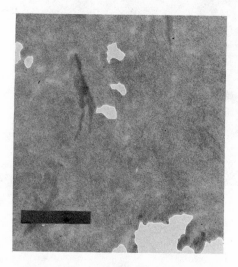

Fig. 12.   Film of MSB 3 cast from chlorobenzene solution at 150°C.
Marker - 0.5μ.

It seems probable that this behavior results from the increasing chain length increasing the viscosity and thereby enhancing the difficulty of the various kinds of blocks to aggregate into the equilibrium microstructure. To test this hypothesis films were cast from diphenylether at 260°C and the solvent removed in a vacuum. One such film of MSB 1 is displayed in Fig. 13. This electronmicrograph should be compared to Fig. 10. Here it is seen that casting from the high temperature solvent does indeed produce structures of higher contrast and hence of presumably greater order. These structures resemble those in the ABA block copolymers. Further studies, casting from diphenylether at 260°C, showed that the size of the dark crystalline domains is related directly to the length of the diphenylsiloxane block.

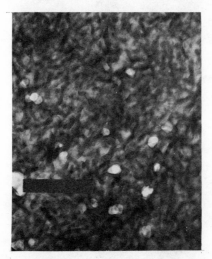

Fig. 13. Film of MSB 1 cast from diphenyl ether at 260°C. Marker - 0.5μ.

The effect of diluting MSB 1 with polydimethylsiloxane of DP 2400 is shown in Figs. 14 and 15. These should be compared to Fig. 10 which shows the undiluted multisequence block copolymers. It is seen that as dilution proceeds the rodlike entities become more disperse. These dilution studies support the contention that the dark rods are indeed 'diphenyl' crystalline regions.

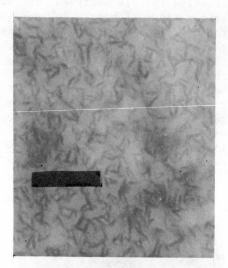

Fig. 14.   Film of 75% MSB 1 plus 25% polydimethylsiloxane cast
from chlorobenzene. Marker - 0.5μ.

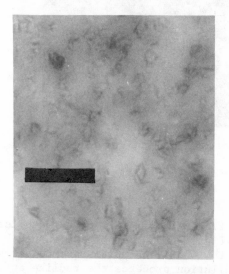

Fig. 15.   Film of 50% MSB 1 plus 50% polydimethylsiloxane cast
from chlorobenzene. Marker - 0.5μ.

CONCLUSIONS

Block copolymers of the ABA type of diphenyl and dimethyl-siloxane when precipitated from dilute solution form lamellae and threads. The lamellae (a few hundred Angstroms thick) are sometimes terraced and the threads, several microns long, are about 200Å in diameter.

Films of these ABA copolymers contain dispersed crystalline regions of the diphenyl block. Annealing of the films above 300°C decreases the size and increases the number and perfection of the crystalline regions. These regions are rods which tend to aggregate with their long axes parallel. The sizes of the crystalline region increase with increasing diphenyl block length.

In the multisequence block copolymers similar morphology prevails. However, the increased molecular length tends to diminish the rate of attainment of these morphologies.

REFERENCES

1.   H. Benoit, P. Rempp, and E. Franta, Compt. Rend., 257(6), 1288-90 (1963).

2(a) K. Kato, Japan Plastics, p. 6, April 1968.

 (b) K. Kato, Polymer Letters 4, 35, (1966).

 (c) H. Hendus, K. H. Illers, and E. Ropte, Kolloid-Z. u.Z.f. Polymere, 216/217, 110 (1967).

3.   T. Inoue, T. Soen, T. Hashimoto and H. Kawai, J. Polymer Sci., A-2 (in press).

4.   E. E. Bostic, This volume p. 237.

STRUCTURE AND PROPERTIES OF ALTERNATING BLOCK POLYMERS OF DIMETHYL-

SILOXANE AND BISPHENOL-A CARBONATE

Roger P. Kambour

General Electric Research and Development Center

Schenectady, New York

Evidence from several kinds of studies indicates that in many block polymers in the solid state a compositional fluctuation or so-called domain structure exists on the scale of 10^2 A ca.[1-6] Many physical properties of these systems are well known to differ from those of their random-copolymerized counterparts because of this structure. Block polymers of dimethylsiloxane and bisphenol-A carbonate have been synthesized by H. A. Vaughn[7] which exhibit physical properties reminiscent of those of other block polymers composed of at least one rubbery block separating two glassy blocks.

I. SYNTHESIS

Each material may be thought of as resulting from a copoly-condensation between an α,ω-dichlorodimethylsiloxane oligomer and a bisphenol-A carbonate oligomer. The synthesis is straightforward: A preformed α,ω-dichloropoly(dimethylsiloxane) (I) is allowed to react with excess bisphenol-A in a pyridine-methylene chloride solution to form II. Phosgene and additional bisphenol-A are added to II to form polycarbonate blocks in situ. After removal of pyridine and pyridine hydrochloride the copolymer is isolated by solvent evaporation or precipitation in methanol. The materials are alternating, random block polymers of the -ABABA---type in which the blocks are polydisperse. Polymer weight average molecular weight lies in the range 50,000-100,000 ($[\eta] \simeq 1.0$ in chloroform).

In these equations n and m correspond to number average molecular weights and are calculated in each case from end group analysis on I and the overall composition of the polymer.

263

$$Cl\left[\begin{array}{c}CH_3\\|\\Si-O\\|\\CH_3\end{array}\right]_{\bar{n}-1}\begin{array}{c}CH_3\\|\\Si-Cl\\|\\CH_3\end{array} + 2HO\text{—}\bigcirc\text{—}\begin{array}{c}CH_3\\|\\C\\|\\CH_3\end{array}\text{—}\bigcirc\text{—}OH$$

EXCESS PYRIDINE

I   (1)

$$HO\text{—}\bigcirc\text{—}\begin{array}{c}CH_3\\|\\C\\|\\CH_3\end{array}\text{—}\bigcirc\text{—}O\left[\begin{array}{c}CH_3\\|\\Si-O\\|\\CH_3\end{array}\right]_{\bar{n}}\bigcirc\text{—}\begin{array}{c}CH_3\\|\\C\\|\\CH_3\end{array}\text{—}\bigcirc\text{—}OH$$

II

$$II + (\bar{m}-2)\ HO\text{—}\bigcirc\text{—}\begin{array}{c}CH_3\\|\\C\\|\\CH_3\end{array}\text{—}\bigcirc\text{—}OH + (\bar{m}-1)\ COCl_2$$

EXCESS PYRIDINE   (2)

$$\text{www}\left[\left(\begin{array}{c}CH_3\\|\\Si-O\\|\\CH_3\end{array}\right)_{\bar{n}}\bigcirc\text{—}\begin{array}{c}CH_3\\|\\C\\|\\CH_3\end{array}\text{—}\bigcirc\text{—}O\left(\begin{array}{c}O\\||\\C-O\end{array}\text{—}\bigcirc\text{—}\begin{array}{c}CH_3\\|\\C\\|\\CH_3\end{array}\text{—}\bigcirc\text{—}O\right)_{\bar{m}-1}\right]\text{wwww}$$

When cast from methylene chloride solutions, these materials form clear films varying in character from strong rubbers to tough plastics depending on bisphenol-A carbonate (BPAC) content. Instron load-elongation curves for several materials of constant average siloxane block length ($\bar{n} = 20$) are shown in Fig. 1.[7] At low BPAC content ultimate elongation is high but largely elastic. With increasing BPAC content a yield stress is developed and recovery after failure decreases. At high BPAC content yielding involves neck formation as it does in homopoly(bisphenol-A carbonate). Fig. 2 shows tensile strengths and elongations at break vs. BPAC content for two series of polymers of $\bar{n} = 20$ and 40, respectively.[7]

II. TORSION PENDULUM STUDIES

Using a simple torsion pendulum described elsewhere[8] the shear modulus G' and the logarithmic decrement $\triangle$ were determined on cast films of several materials from -150°C to the temperature at which creep under the weight of the torsion bar became too great.[9] Frequencies were in the range 0.3 to 3 Hz.

Fig. 3 shows log G' and $\triangle$ vs. T from -150°C to +90°C for one material (51% BPAC; $\bar{n} = 20$, $\bar{m} = 6$). Two major relaxations are evident: one at about -110°C and one at 72°C. Figure 4 shows log G' behavior in the low temperature range for homopoly(bis-

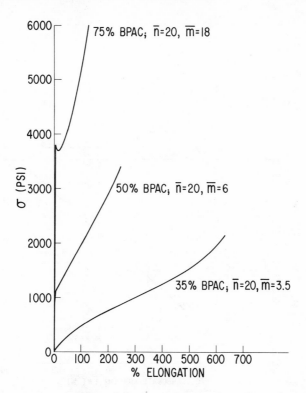

Figure 1. Engineering stress (σ) vs. elongation for three dimethyl-siloxane-bisphenol-A carbonate block polymers of the compositions indicated. Instron crosshead speed = 2"/min.

phenol-A carbonate) and several block polymers of varying block length and composition. In general, the size of the relaxation increases with both decreasing BPAC content and increasing siloxane block length. Fig. 5 shows log G' behavior of one block polymer (35% BPAC) compared to that of a lightly crosslinked specimen of unfilled polydimethylsiloxane.

Although polydimethylsiloxane crystallized partially, its T_g (about -123°C)[10] is still clearly reflected in both mechanical and dielectric[11] relaxation curves. In the block polymers studied to date ($\bar{n}$ = 5 to 20) no evidence of a crystal melting "hump" has been seen in the G' vs. T behavior. Furthermore, although homopoly(bis-phenol-A carbonate) exhibits a very broad loss peak at about -100°C,[12] the trends in the size of the drop in log G' with increasing siloxane content evident in Fig. 4 indicate that in the copolymers relaxation in the -100°C range is associated with the T_g of siloxane blocks primarily.

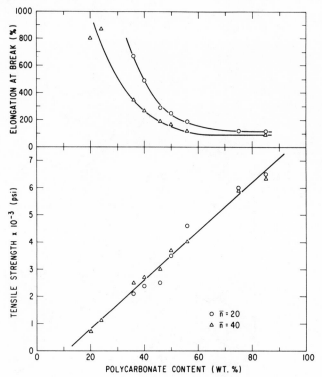

Figure 2.   Tensile strengths and ultimate elongations of two block polymer series of $\bar{n}$ = 20 and 40, respectively.

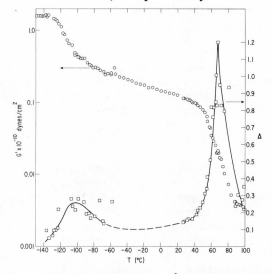

Figure 3.   Dependence of shear modulus G' and logarithmic decrement $\triangle$ on temperature at about 1 Hz for one block polymer (51 wt. % BPAC; $\bar{n}$ = 20, $\bar{m}$ = 6).

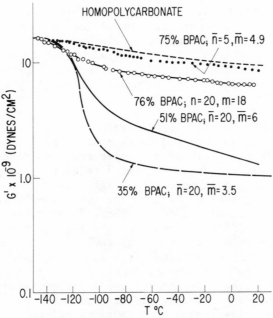

Figure 4. Shear modulus G' in low temperature region for homopoly-(bisphenol-A carbonate) and several block polymers of indicated compositions. Each curve shifted on log G' scale to bring it into registry with that of homopolycarbonate at -150°C.

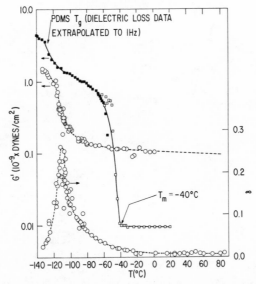

Figure 5. Shear modulus G' in low temperature region for homopoly-(dimethylsiloxane) (squares) and one block polymer (circles) of 35% BPAC ($\bar{n} = 20$, $\bar{m} = 3.5$). Logarithmic decrement of copolymer also shown.

The size of the log G' drop in Fig. 4 is seen to depend also on siloxane block length at constant composition which presumably reflects the fact that the polycarbonate blocks act as virtual crosslinks among other things. The relatively small dependence of the temperature of this relaxation on block length is probably attributable to the extremely high segmental flexibility of the dimethylsiloxane unit.[11]

Fig. 6 shows the upper temperature relaxation behavior for several of these materials compared to that of homopoly(bisphenol-A carbonate).[12] Here both relaxation temperature and size of the drop in log G' are seen to vary. In Fig. 7 relaxation temperature is plotted against reciprocal molecular weight of these BPAC blocks. Also included are α relaxations at 1 Hz for two homopolycarbonates estimated from stress relaxation data[13] and T_g's by differential thermal analysis at a 2.5°C/min. heating rate for a set of homopolycarbonates of a wide molecular weight range.[14] All mechanical relaxation data are seen to be fitted by a smooth descending curve. The calorimetric results on the homopolymers lie on a single line of lesser slope. Qualitatively these lines are reminiscent of the straight line obtained on a similar plot by Fox and Flory[15] for dilatometric T_g's of polystyrenes of a range of molecular weights.

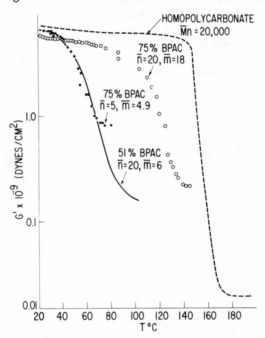

Figure 6. Upper temperature relaxation behavior of several block polymers compared to homopolycarbonate. Curves have been shifted vertically on log G' scale to bring them all into approximate registry with homopolycarbonate at room temperature.

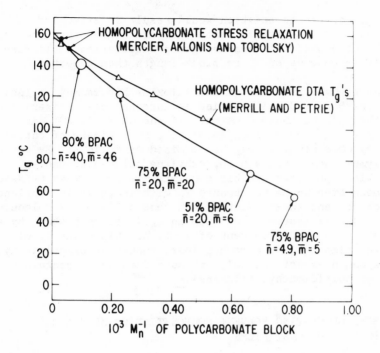

Figure 7. Dependence of upper relaxation temperature on reciprocal $\overline{M}_n$ of BPAC block (o) and comparison with α relaxation temperatures of two homopolycarbonates (●) and DSC T_g's of a series of homopoly-carbonates (△).

These data suggest that the upper mechanical relaxation in the block polymers is a manifestation of the T_g's of polycarbonate domains. The molecular weight dependence of the relaxation tempera-tures is largely the analogue of the dependence of T_g on molecular weight in homopolymers. Presumably the siloxane blocks attached to the ends of each BPAC block are so mobile at these temperatures[11] that they exert little rotational restraint on the polycarbonate blocks which thus behave largely as though they had free ends.

The discrepancy between the mechanical relaxation temperatures of the copolymers and the differential thermal analysis T_g's of the homopolymers is 15-20°C at low molecular weights. This difference may arise from one or more factors. First, as a matter of experi-ence, the comparison of T_g's obtained by methods as diverse as mechanical relaxation and differential thermal analysis often shows significant discrepancies which are likely to be the greater where different sets of workers are also involved. Second, T_g is likely to be somewhat affected by the chemical nature of the end group; the homopolycarbonates cited here have hydroxyl end groups which may well be associated with each other through hydrogen

bonding. If so, the association would result in effective
molecular weights in the bulk state that are higher than the values
for unassociated chains. Thus end group association would result
in a lesser dependence of T_g on block length than expected.

Third, a small fraction of the siloxane blocks may be included
in the polycarbonate domains acting as internal plasticizers and
lowering the T_g's of these domains. If the whole 15-20° discrepancy is attributed to such an effect, an estimate of the siloxane impurity concentration may be obtained as follows: By contrast
with copolymers under study here, ethylene oxide-bisphenol-A
carbonate block polymers exhibit complete miscibility as evidenced
by the smooth drop in polycarbonate T_g with increase in ethylene
oxide block content independent of ethylene oxide block molecular
weight.[14] In this system a 20° drop in T_g is brought about by a
polyethylene oxide block content of 4 wt. %. Since the T_g of
poly(dimethylsiloxane) is somewhat lower than that of poly(ethylene
oxide), we would expect the polycarbonate domains to contain no
more than 4% poly(dimethyl siloxane).

III. EFFECTS OF CASTING SOLVENT COMPOSITION ON MECHANICAL
PROPERTIES OF CAST FILMS

As with other phase-separated copolymers, films of the dimethyl siloxane-bisphenol-A carbonate block polymers can exhibit
changes in mechanical behavior when the film casting medium is
changed from one which is a nonpreferential solvent to one which
is preferential for one of the blocks. Effects of this kind are
understood to arise from changes in the extent to which the
respective domains form continuous networks. Since the change in
relative degree of network formation with change in polymer composition in a series of 2-phase materials is most striking in the
50% composition range (i.e. the so-called phase inversion range),
preferential solvent effects in cast films are expected to be most
striking for block polymers having compositions in the same range.

Figure 8 shows stress-strain curves of three films of a
single polymer (50% BPAC; $\bar{n} = 20$, $\bar{m} = 6$) that were cast respectively
from solutions of 10% polymer in pure methylene chloride and in
two solvent mixtures having initial methylene chloride/n-hexane
volume ratios of 1:1 and 1:2. The films were evaporated to complete dryness at room temperature in each case. The markedly
greater "rubbery" character to the films cast from the solvent
mixtures is associated with the fact that n-hexane is a nonsolvent
for polycarbonate but a good solvent for poly(dimethyl siloxane).
These effects are displayed most dramatically in a plot of yield
stress (i.e. the stress at the "knee" in the stress-strain curve)
against casting solvent composition (Fig. 9). The drop in yield
stress results from an increasing tendency of the polycarbonate

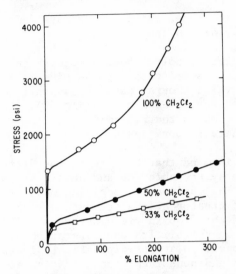

Figure 8. Instron stress-strain curves of films of one block polymer (51 % BPAC; $\bar{n}$ = 20, $\bar{m}$ = 6) cast from methylene chloride and two mixtures of methylene chloride and n-hexane.

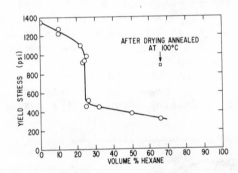

Figure 9. Yield stress of cast films vs. composition of casting solvent.

blocks to coacervate early in the drying process while the siloxane blocks remain extended in the solvent medium. Presumably the abrupt drop at 24% hexane is caused by the occurrence of polycarbonate coacervation at such an early stage in the solvent evaporation that each resultant polycarbonate microdomain is almost completely separated from its neighbors by a "sheath" composed of siloxane blocks which are still heavily swollen by the medium. Subsequently in the resultant dry films there no longer exists a continuous network of polycarbonate and the stress-strain curve is rather like that for a block polymer of low polycarbonate content wherein, for

reasons of volume ratio alone, the polycarbonate domains are
expected to be largely discontinuous.

That the constraint on polycarbonate network formation is
reversible in character is indicated by the rise in yield stress
toward a more normal value brought about by annealing one of the
rubbery films at 100°C for several hours (Fig. 9). Above the T_g
of the polycarbonate domains the reformation of the normal
continuous polycarbonate network again becomes possible.

It must be kept in mind that because of the difference in
vapor pressures of methylene chloride and hexane at room temperature (400 vs. 150 mm. Hg) solvent composition moves from its
original value toward greater hexane contents during evaporation.
Thus the sharpness of the drop in yield stress at 25% hexane
probably is enhanced by the simultaneous operation of two factors
which drive polycarbonate toward precipitation: increase in both
polymer concentration and nonsolvent concentration as evaporation
proceeds.

Another more fundamental indicator of the increasing tendency
for internal coacervation of the polycarbonate blocks in each
macromolecule is afforded by measurements of intrinsic viscosity
of the polymer in various methylene chloride/n-hexane mixtures
(Fig. 10). The break in $[\eta]$ at 20% hexane is not as dramatic as
the drop in yield stress at 24% original hexane content of the
casting medium. That the drop in $[\eta]$ begins at 20% hexane suggests
that polycarbonate block coacervation in the casting solutions of
24% hexane already exists to a degree at the original polymer
concentration or begins to develop shortly after evaporation
begins.

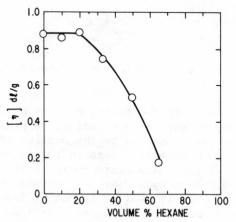

Figure 10. Intrinsic viscosity $[\eta]$ of the same block polymer in
methylene chloride/n-hexane mixtures.

IV.   X-RAY SCATTERING AND ELECTRON MICROSCOPY

Joint studies[9,16] with D. G. LeGrand employing a Kratky slit camera showed that each material exhibited pronounced low-angle x-ray scattering with an interference peak.   In one material (35 wt. % BPAC; $\bar{n}$ = 20, $\bar{m}$ = 3.5) this peak occurred at a Bragg distance of about 150 Å (Fig. 11).   Immersed in n-hexane, a solvent for silicone but a nonsolvent for homopolycarbonate, this material swelled to eight times its original volume and simultaneously the interference peak became more pronounced and shifted to smaller scattering angles. Taken with the 20% greater electron density of polycarbonate relative to that of poly(dimethylsiloxane), the above observations may be interpreted to indicate that the interference peak arises from scattering between polycarbonate aggregates the average center-to-center distance of which is of the order of 100-200 Å in the unswollen film.

Figs. 12a and b are transmission electron micrographs of these films of two copolymers (35 wt. % BPAC: $\bar{n}$ = 40, $\bar{m}$ = 6.3 and $\bar{n}$ = 100, $\bar{m}$ = 15.8, respectively).   These were cast from 1% solutions in CH_2Cl_2.   Micrographs of this type suggest a tendency of the polycarbonate blocks to aggregate in discrete but irregularly shaped "domains" separated by siloxane-rich regions.

According to Williams and Flory[17] the poly(bisphenol-A carbonate) chain repeat unit may be represented by a succession of virtual bonds 7.0 Å in length, joined at angles of ca. 112°.   Thus each BPAC unit adds 2 x 7 x sin (112°/2) = 11.6 Å to the contour chain length. Consequently, for $\bar{m}$ = 6.3 and 15.8 the number average contour block lengths are 73 Å and 184 Å, respectively.   These lengths and the sizes of the electron density fluctuations in Figs. 12a and b are roughly the same.   Thus when $\bar{m}$ is reasonably small, the domain diameter is about equal to the contour length of the number average BPAC block.

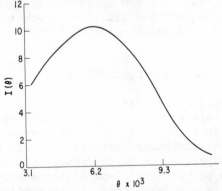

Figure 11.   Low-angle x-ray scattering from one block polymer (35 wt. % BPAC; $\bar{n}$ = 20, $\bar{m}$ = 3.5) showing interference peak at Bragg distance of 150 Å.   Copper Kα radiation.

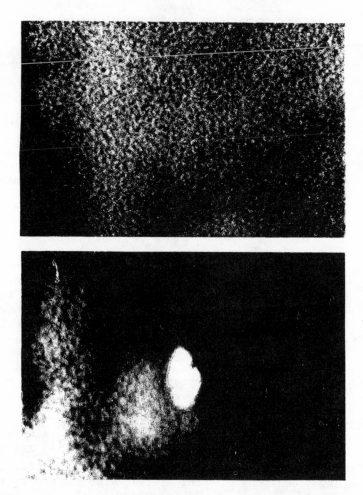

Figure 12.   Transmission electron micrographs of thin films of two
copolymers of 35 wt. % BPAC but different block lengths:
(a) $\bar{n}$ = 40, $\bar{m}$ = 6.3; and (b) $\bar{n}$ = 100, $\bar{m}$ = 15.8.

V.   CONCLUSIONS

On the basis of these studies it appears that as in other
block polymers the bulk properties of these materials are successfully rationalized in terms of a microdomain structure in each
material.   At low polycarbonate contents discrete particles composed largely of polycarbonate blocks exist, at intermediate
concentrations both polycarbonate and silicone continua exist and
at high polycarbonate concentrations discrete silicone aggregates
exist in a polycarbonate continuum.   Both microphases give rise to
relaxations closely related to the T_g's of the corresponding homopolymers.   On the basis of comparison with the behavior of the two

corresponding homopolymers and a series of polycarbonate-poly-ethylene oxide block polymers, it appears that in the materials under study here the polycarbonate domains are largely free of siloxane block "impurities" and vice versa.

By contrast with the ternary block polymers of styrene and butadiene in which microdomains are thought to exist only for $M_{PS} \geq 5000$ ca.,[18] these features appear to persist down to poly-carbonate block molecular weights averaging less than 875. This difference undoubtedly reflects the much larger disparity in solubility parameters here (7.5 and 10) than in the case of poly-butadiene and polystyrene (8.4 and 8.8).[19]

REFERENCES

1. A. Skoulios, G. Tsouladze and E. Franta, J. Polymer Sci., C4, 507 (1963).

2. M. Baer, J. Polymer Sci., A2, 417 (1964).

3. E. Vanzo, J. Polymer Sci., Part A-1, 4, 1727 (1966).

4. S. L. Cooper and A. V. Tobolsky, J. Applied Polymer Sci., 10, 1837 (1966).

5. H. Hendus, K.-H. Illers and E. Ropte, Kolloid-Z., 216-217, 110 (1967).

6. J. F. Henderson, K. H. Grundy and E. Fischer, J. Polymer Sci., Part C, 16, 3121 (1968).

7. H. A. Vaughn, J. Polymer Sci., Part B, 7, 569 (1969).

8. R. P. Kambour, F. E. Karasz and J. H. Daane, J. Polymer Sci., Part A-2, 4, 327 (1966).

9. R. P. Kambour, J. Polymer Sci., Part B, 7, 573 (1969).

10. C. E. Weir, W. H. Leser, and L. A. Wood, J. Research Nat'l. Bur. Standards, 44, 367 (1950).

11. C. M. Huggins, L. E. St. Pierre and A. M. Bueche, J. Phys. Chem., 64, 1304 (1960).

12. D. G. LeGrand and P. F. Erhardt, J. Applied Polymer Sci. (1969), in press.

13. J. P. Mercier, J. J. Aklonis, M. Litt and A. V. Tobolsky, J. Applied Polymer Sci., 9, 447 (1965).

14. S. H. Merrill and S. E. Petrie, J. Polymer Sci., Part A, <u>3</u>, 2189 (1965).

15. T. G. Fox and P. J. Flory, J. Polymer Sci., <u>14</u>, 315 (1954).

16. D. G. LeGrand, J. Polymer Sci., Part B, <u>7</u>, 579 (1969).

17. A. D. Williams and P. J. Flory, J. Polymer Sci., Part A-2, <u>6</u>, 1945 (1968).

18. D. J. Meier, J. Polymer Sci., Part C, <u>26</u>, 18 (1969).

19. <u>Polymer Handbook</u>, J. Brandrup and E. H. Immergut, Editors, Interscience Publishers, New York, 1966.

PREPARATION OF BLOCK COPOLYMERS OF SOME

ETHYLENIC MONOMERS AND PROPYLENE SULFIDE

Albert GOURDENNE

Laboratoire de Chimie Macromoléculaire de la

Faculté des Sciences de Paris, 1 rue V.Cousin

PARIS V

INTRODUCTION

The anionic polymerization is an addition polymerization, where the active centers, called living ends
by Professor Szwarc (1,2), are carbanions. They are
negatively charged species ; however, since the solution is electrically neutral, positive counterions
must be present in an equivalent amount. Most of the
investigated polymerizations take place in media
having relatively low dielectric constants ; therefore,
an association of the oppositly charged ions into ion
pairs generally occurs (3). There is an equilibrium
between ions and ion pairs :

$$\sim\!\!\sim\!\!\sim\!S^-,Me^+ \quad \rightleftharpoons \quad \sim\!\!\sim\!\!\sim\!S^- \quad + \quad Me^+ \qquad (1)$$

The anionic polymerization is now certainly the
best method to prepare homopolymers and copolymers
having a Poisson molecular weight distribution. Its
success is based on the assumption that all the polymer chains grow simultaneously with an equal rate and
none is terminated before the completion of the
preparation (4). The relative stability of the living
polymers secures the latter conditions. But, when free
ions and ion pairs participate in the process, all the
active centers have not the same reactivity ; free
ions are much more reactive than ion pairs. Thus not
all the polymers grow with the same rate. However,
there is an exchange between the ions and the ion

277

pairs. The ion pairs can dissociate into ions, whereas
the ions can associate into pairs, and therefore the
molecular weight distribution depends on the rate of
the exchange. Generally the rate of exchange is faster
than the rate of growth and the molecular weight distribution of the resulting polymer is not different
from Poisson distribution. All our results assume that
fact. In order to avoid any hindered exchange due to
the viscosity of the medium, polymerization reactions
have to be carried in dilute solution.

The stability of the living ends is the main
condition required for obtaining well defined block
copolymers. This important point is discussed now.

STABILITY OF THE ACTIVE CENTERS OF SOME LIVING POLYMERS

Two kinds of initiators give carbanions when
they react with ethylenic or heterocyclic monomers,
bases such as butyllithium, and adducts of metals
with polycyclic hydrocarbons such as sodium biphenyl.
In the first step, the base fixes the monomer whereas
the adduct initiates the reaction by electron transfer.
Very often the initiation reactions are complicated.
Then, the mechanism of propagation is theoretically
simple :

$$\sim\!\!\sim\!\!C \ Me \quad + \quad M \quad \longrightarrow \quad \sim\!\!\sim\!\!C \ M \ Me \qquad (2)$$

where $\sim\!\!C$Me is the growing chain and M the monomer.
When the monomer is added, the growth can continue on
condition that the active centers remain stable.

The stability of the living ends depends on the
nature of the monomer to be polymerized, the solvent
and the initiator, and on the experimental conditions
of polymerization, i.e. temperature and concentration
of reagents.

When the carbanions are not stable, two types of
reactions can occur : the "isomerization" which is
only a change of the nature of the chemical species
without transfer reaction, and the deactivation which
involves a transfer.

1.- Reactions of Isomerization of Living Ends

When initiation of isoprene is done in tetrahy-
dorfuran (THF) by a sodium film at room temperature,
the U.V. spectrum gives only a peak with a maximum

situated at 358 nm (5). When initiation is done at low temperature, at -40°C for example, a new peak can be seen at 308 nm. Then it is displaced slowly with time towards higher wavelenghts until at 358 nm where it is stabilized, with a decrease of its intensity (6). There is probably no destruction by transfer reactions at that temperature where their activation energies are weak. The last form absorbing at 358 nm is called "isomerized" form. It appears immediatly when initiation takes place at room temperature with small amount of monomer, whereas, when the initiation is done at low temperature, the initial form absorbing at 308 nm can be seen before its change into the isomerized form. This transformation could be explained by Bywater (5), who thinks that in case of living polyisoprene in THF a proton of the CH3 group migrates to give a more symmetrical carbanion or isomerized carbanion. But this explanation is not convenient for understanding the mechanism of isomerization of sodium polybutadienyl in THF at room temperature, and which is quite stable in intensity and wavelenght at low temperature. In fact, in the case of polybutadiene, there is no CH3 group.

The isomerized forms of sodium polybutadienyl and sodium polyisoprenyl are active for the polymerization of each diene respectively; the initial species are reformed after new addition of monomer but are isomerized again after the end of the polymerization (6). Therefore the isomerization is a rearrangement of the organic part of the initial carbanion, without transfer reaction and which cannot always be avoided by working at low temperature. The reactivity of the isomerized form is probably different from that of the initial species.

2.- Reactions of Deactivation of Living Ends

The deactivation of the carbanions occurs also very often. There is a disappearance of the carbanion by transfer reactions which can be intramolecular or intermolecular.

In the cases of sodium polystyryl (7) and sodium polybutadienyl (8), at room temperature, the living polymers are deactivated by removal of sodium hydride and there is formation of a double bond. Particularly in the case of polybutadiene, the new double bond is conjugated with an other one and its reactivity increases towards organometallic compounds contained in the solution. It seems that the removal of sodium hydride is a general phenomenon of intramolecular deactivation. However it is not always the main mechanism of deactivation. For example, in the case of sodium polyparadimethylaminostyrene, the attack of the carbanion on dimethylanilin groups is faster than removal of sodium hydride (9). This attack on dimethyl anilin groups is a reaction which can be as well intramolecular as intermolecular.

The carbanions react with impurities contained in the polymerization solution. For example, in the case of initiation of the polymerization of dienes by adducts of sodium with polycyclic hydrocarbons, when they are not well purified, the living ends react even at low temperature with dihydrogenated forms of the polycyclic hydrocarbons. Thus the polymerization of dienes is stopped, and the new metallated product issued from deactivation can initiate the polymerization of weak monomers such as episulfides; so that reactions of copolymerization can lead to polydisperse copolymers with large amounts of homopolymers (6).

The solvent has also a great influence. For example, the species lithium polyisoprenyl and lithium polybutadienyl are quite stable in cyclohexane solution, but not in THF (5), where the stability decreases by transfer reaction to the solvent (10).

3.- Other Reactions Changing the Nature of Living Ends

The products issued from the reactions of deactivation of the carbanions can react with the non deactivated carbanions. For example, in the case of sodium polystyryl in THF (7), the living ends (I) are destroyed by sodium hydride removal, with formation of a double bond. The ethylenic product which is obtained (II), can react with the carbanion of polystyrene (I) to give a new carbanion (III), unreactive for the polymerization of styrene :

$$\sim\sim CH_2 \overset{(-)}{-} \overset{(+)}{CH} \; Na \qquad \longrightarrow \qquad \sim\sim CH = CH \qquad NaH \qquad (3)$$

$$\underset{(I)}{\underset{\varphi}{|}} \qquad\qquad\qquad \underset{(II)}{\underset{\varphi}{|}} \; +$$

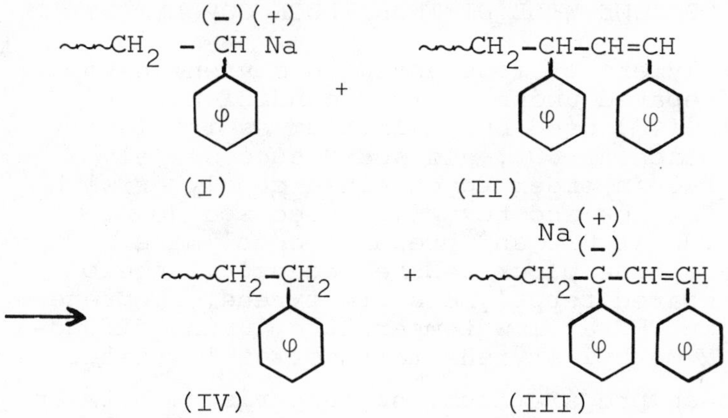

$$
\begin{array}{cc}
\overset{(-)}{\sim\sim CH_2} - \overset{(+)}{CH}\ Na & \sim\sim CH_2-CH-CH=CH \\
\varphi & \varphi \qquad \varphi \\
(I) & (II)
\end{array}
$$

$$
\sim\sim CH_2-CH_2 \quad + \quad \sim\sim CH_2-\overset{Na^{(+)}}{\underset{(-)}{C}}-CH=CH \qquad\qquad (4)
$$

$$
\begin{array}{cc}
\varphi & \varphi \qquad \varphi \\
(IV) & (III)
\end{array}
$$

Some other reactions can change the nature of living ends and stop the propagation process. For example, it is well known that anthracene can be added to living ends of polystyrene (I) [11]. It is a 9-10 addition of the polycyclic hydrocarbon (II) to the active center :

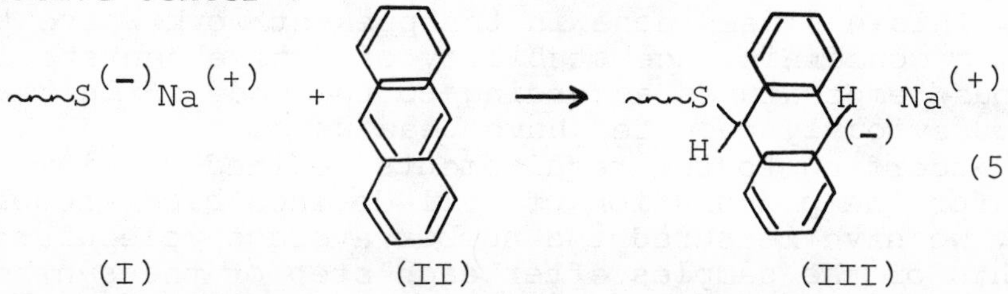

$$
\sim\sim S^{(-)} Na^{(+)} \quad + \qquad \longrightarrow \qquad \sim\sim S \qquad H \quad Na^{(+)}_{(-)} \qquad (5)
$$

$$
\begin{array}{ccc}
(I) & (II) & (III)
\end{array}
$$

The new product (III) is unreactive towards styrene. When the initiation of the polymerization of dienes is done by monosodium anthracene, which involves an electron transfer, free anthracene reacts immediatly after the initiation with living ends to give a metal-lated anthracenated compound which has no reactivity towards dienes (12).

Many reactions of deactivation and isomerization can be avoided under defined conditions of purity of reactants and of temperature. They also take place during the copolymerization reactions and give either no copolymers or copolymers with complicated structure (13). Thus the possible reactions of every living end in polymerization solution have to be studied in order to obtain well defined block copolymers.

PREPARATION OF SOME WELL DEFINED BLOCK COPOLYMERS

Block copolymers of isoprene with styrene have
already been prepared under various conditions.
KOROTKOV et al. (14) used butyl lithium as initiator
in benzene solution at 60°, and added successively
the two monomers, in order to obtain a copolymer with
three blocks. Schlick and Levy (15) used sodium
naphtalene in THF at 0°C and prepared copolymers
believed to have nine blocks. More recently, Angelo
et al. (16) prepared terpolymers of styrene, isoprene
and butadiene in THF at low temperature, using difunctional living α methyl styrene tetramer as initiator.

In all these preparations, no comparisons between
theoretical molecular weights and experimental molecular weights of various polymers were made. It is likely
that copolymers were obtained, but there is no proof of
the supposed structure. In view of the low stability of
living ends of dienes, a comparison between experimental and theoretical molecular weights would have had
to be made together with a verification of the nature
and the stability of carbanions during the polymerization. This has been done in the present work where the
results concerning the stability of active centers at
various temperatures according to the mode of initiation previously reported have been used.

According to the requirements defined by Sigwalt
(17) for the preparation of well-defined block copolymers, we have measured the number average molecular
weights of the samples after each step of the synthesis,
and also the weight average molecular weights in order
to follow the variation in polydispersity due to each
sequence.

Various molecular weights have to be defined. The
theoretical number average molecular weight, $\overline{M}_n$th, is
calculated from the total quantity of polymer obtained
after evaporation of the solvent and from the spectros-
copic data giving the initiator concentration. The
experimental number average molecular weight, $\overline{M}_n$osm.,
is determined by osmotic pressure measurement. The
weight average molecular weight, $\overline{M}_w$, is obtained by
light scattering (Debye's method) and it is necessary
to use three solvents in the case of copolymers (18).
The calculated number average molecular weight, $\overline{M}_n$calc.,
is calculated only for the copolymers from experimental
$\overline{M}_n$ osm. of the homopolymers and the quantities of
other monomers added. This value, $\overline{M}_n$calc., accounts
for the possible deactivation of carbanions after

sampling and copolymerization reactions.

In the case of initiation of polymerization by strong bases such as butyl lithium, the first block to be prepared A is a monofunctional living polymer. When the second monomer is added, a copolymer with two blocks symbolized A B, is synthetized. If the initiation is done by adducts of sodium with polycyclic hydrocarbons i.e. by electron transfer, the first block A is bifunctional. By addition of a second monomer, a copolymer with three blocks is obtained and can be symbolized BAB. In this way, a copolymer with five blocks between isoprene and butadiene can be written $A_2BA_1BA_2$, A is polyisoprene and B is polybutadiene.

The synthesis of well defined block copolymers of isoprene and butadiene on the one hand, and of some ethylenic monomers and episulfides on the other hand, is now possible and will be now described.

1/- Copolymers of Isoprene and Butadiene

a)- Copolymer of Isoprene and Butadiene with two blocks (6, 19)

The aim of this synthesis is to prepare a copolymer having one sequence of 1,4 cis polyisoprene and one sequence of 1,2 polybutadiene. The first sequence was obtained in hexane by using seeds of lithium polyisoprenyl (20). The structure of the homopolymer in this case is essentially 1,4 cis (21). Then, butadiene in THF solution was added at -4°C, the carbanion of the living polyisoprene initiated immediatly the polymerization of butadiene, THF destroyed the associations of lithium polyisoprenyl (22, 23) and gave a 1-2 structure for the polybutadiene (24). It was necessary to work at low temperature to avoid the deactivation of every type of living ends in solution. During the polymerization, the position and the intensity of the maximum of the peak in the UV spectrum, corresponding to the absorption of lithium polybutadiene did not change. This is in agreement with the results on the stability of living polybutadiene previously reported (6).

The results of the molecular weights measurements are given in Table 1.

Table 1. Preparation of a Block Copolymer of Isoprene
         and butadiene with two blocks.

Polymer	$\overline{M}_n$ th.	$\overline{M}_n$ calc.	$\overline{M}_n$ osm.	$\overline{M}_w$	$\overline{M}_w/\overline{M}_n$ osm.
Homopolymer	115000	–	138000	162000	1,17
Copolymer	237500	288000	330000	345000	1,04

For the homopolymer, the agreement between
$\overline{M}_n$ osm and $\overline{M}_n$ th is not good. Similar results have
been found for different homopolymers using the seeds
technics (20). In the case of the copolymer, $\overline{M}_n$ osm.
is not quite different from $\overline{M}_n$ calc. ($\overline{M}_n$ osm./ $\overline{M}_n$ calc.
~ 1,1). Therefore the deactivation does not seem to
be important and the synthesis of the copolymer, well
established by the general increase of various molecular weights, gives a monodisperse copolymer ($\overline{M}_w$/
$\overline{M}_n$ osm. ~ 1,04).

b)- Copolymers of isoprene and butadiene with
    three and five blocks (6, 19)

In THF, the fundamental carbanions of sodium
polybutadienyl (λ = 293 nm) and sodium polyisoprenyl
(λ = 308 nm) and the corresponding isomerized forms
(λ = 384 nm for polybutadiene and λ = 358 nm for
polyisoprene), are active for the polymerization of
butadiene and isoprene. Thus, when the polymerization
is done at low temperature to avoid deactivation and
when the purity of reactants is very good, it is
always possible to obtain block copolymers. However,
it is necessary to consider many factors : the temperature of the experiment, the concentration of reactants and the selected times of various steps of the
synthesis. In fact, the structure of each block, in
particular in the case of polybutadiene, depends on
the temperature (24). The rate of isomerization of
initial carbanions, which cannot be avoided after the
end of polymerization of isoprene in the case of
sodium counter ion (6), depends also on all these
factors. The isomerized forms have probably a reactivity different from that of initial species, which
can introduce polydispersity and also give an other
structure for the polybutadiene block if butadiene is
added on isomerized forms of living polyisoprene
instead of fundamental forms. A copolymer is well defined when the conditions of synthesis are well known

and the molecular weights determined.

The synthesis of a five blocks copolymer is possible. Many details are now being given to describe one of our typical experiments.

The initiator (6,40 . 10^{-4} mole/l) was sodium biphenyl, purified on living polybutadiene. Thus the solution (225 ml) contained very few hydrogenated impurities (6). Isoprene (3,07 g) was polymerized during 4 hours at -40°C and one hour at -72°C. This last step at -72°C was necessary for sampling. Then butadiene (3,84 g) was added at -40°C. The solution stayed 2 hours at -40°C and one hour at -72°C. The new addition of isoprene (1,53 g) was realised at -40°C. After 2 hours at -40°C, the solution was brought at room temperature for 16 hours. Then the active centers are deactivated by methanol.

The spectra of the solution of the initiator and of the living ends were made just before the addition of each monomer. Thus butadiene had been fixed on a mixture (λ = 350 nm) of fundamental (λ = 308 nm) and isomerized (λ = 358 nm) carbanions of polysioprene. The new addition of isoprene had been done on carbanions of polybutadiene absorbing at 305 nm, whereas the fundamental species absorb at 293 nm. There was certainly in this case a shift towards higher wavelengths due to the UV absorption of free biphenyl (λ = 250 nm) and of polymer chains. The spectrum determined after the last addition of isoprene and the spontaneous deactivation of carbanions (16 hours at room temperature), presented only a very small peak, the maximum of which is situated at 419 nm, and is due to the presence of metallated compounds of dihydrogenated forms of biphenyl in small quantity (6).

Table 2 shows the results of molecular weights measurements, which assure that the synthesis of well defined block copolymers of isoprene and butadiene with three and five sequences is established.

Table 2.- Preparation of a block copolymer of isoprene and butadiene with 5 blocks (3 for polyisoprene and 2 for polybutadiene)

	Molecular weights of the polymer obtained after :		
	First addition of monomer (Isoprene)	Second addition of monomer (Butadiene)	Third addition of monomer (Isoprene)
$\overline{M}_n$ th	42 000	177 000	287 000
$\overline{M}_n$ calc.	-	185 000	293 000
$\overline{M}_n$ osm.	41 500	190 000	303 000
$\overline{M}_w$	46 500	209 000	350 000
$\overline{M}_w/\overline{M}_n$ osm.	1,12	1,10	1,15

2/- Copolymers of some Ethylenic Monomers and Propylene Sulfide.

The episulfides are heterocyclic monomers which can be polymerized by ring opening in anionic way and give living polymers. The active centers which are mercaptide groups, have no UV absorption but seem to be very stable. This last result is deduced from molecular weights measurements of obtained polymers, which are in good agreement with expected values from an ionic mechanism (25).

In order to obtain copolymers where the polysulfides are the last blocks to be prepared, only the initiation by bases, i.e. carbanions of other living polymers, is interesting for these synthesis.

The episulfide which is polymerized here is propylene sulfide, $CH_3-CH-CH_2$. The copolymerization
$\overset{\smallsmile}{S}$
involves the opening of the heterocycle. However, the initiation reactions of this episulfide are often complicated by transfer reactions (26).

Some polymerizations, which lead to the synthesis of copolymers including polypropylene sulfide blocks, are now described.

2/-a. Copolymer of isoprene and propylene sulfide with two blocks (6, 27)

Propylene sulfide is an elastomer with interesting properties, and it might be useful to associate these with the properties of 1,4 cis polyisoprene. As propylene sulfide polymerization is slow in hydrocarbons, particularly in the case of lithium counterion, the copolymer was prepared in a mixed solvent (hexane-THF). The reaction of copolymerization of the two monomers was conducted thus : to the solution of the first block of living polyisoprene (obtained in hexane with lithium counter ion) was added at -78°C the episulfide dissolved in THF. The volume of THF represented 58 per cent of the total volume. When the mixture was made, the solution of polymerization became very lightly yellow. After five minutes, at low temperature, the solution was brought to room temperature and allowed to stand for three days at that temperature. At that time, a first sample was taken (copolymer I). After 10 days, at room temperature and 20 hours at 50°C, the conversion of propylene sulfide was complete (copolymer II). Table 3 gives the results of molecular weights measurements for the various polymers which were prepared.

Table 3.- Preparation of a block copolymer of isoprene and propylene sulfide with 2 blocks.

Polymer	$\overline{M}_n$ th.	$\overline{M}_n$ calc.	$\overline{M}_n$ osm.
Homopolymer	135 000	–	175 000
Copolymer I	192 000	248 000	280 000
Copolymer II	245 000	315 000	350 000

The determination of various average molecular weights shows that the synthesis of the copolymer 1,4 cis polyisoprene - polypropylene sulfide is possible, that the reaction of ring opening prevails on the deactivation reactions, and also that the living ends of polypropylene sulfide which are mercaptide groups, are stable in THF at room temperature in the case of lithium counter ion.

2/-b. Copolymers of Isoprene and Propylene
Sulfide, Butadiene and Propylene Sulfide, 2 Vinyl
Pyridine and Propylene Sulfide, with Three Blocks.

Two types of copolymers including polypropylene
sulfide with three blocks have been already prepared.
The two couples which were used, were propylene sulfi-
de-styrene and propylene sulfide-ethylene sulfide,
and the initiator was sodium naphtalene in THF solution (28).

Some new copolymers of ethylenic monomers and
propylene sulfide have been recently synthetized. The
initiators, sodium naphtalene or sodium biphenyl, have
been purified on sodium polybutadiene.

The initial or isomerized carbanions of polyisoprene and polybutadiene can initiate the polymerization of propylene sulfide at -40° (27) whereas those
of poly 2 vinylpyridine cannot (29). In this last case,
initiation reaction has to be done at room temperature,
where it is still not fast. In order to also obtain
well defined block copolymers of 2 vinylpyridine with
propylene sulfide, it seems to be necessary to initiate the polymerization of 2 vinylpyridine by flow
addition* of a small amount of monomer to the initiator
solution (29).

Every type of copolymer synthetized here contains
less than 4 per cent of homopolymer. Table 4 gives all
the results of average molecular weights measurements,
which establish that copolymerization occurs between
propylene sulfide and the ethylenic monomers which are
used here, i.e. butadiene, isoprene and 2 **vinylpyridine**
The resulting copolymers have also a narrow molecular
weights distribution.

All these synthesis of copolymers of ethylenic
monomers and propylene sulfide with two or three blocks
previously reported are in agreement with the following
mechanism for the initiation of copolymerization (26) :

$$\sim\!\!\sim\!CH_2Na \; + \; CH_2\text{-}CH\text{-}CH_3 \qquad \sim\!\!\sim\!CH_2SNa$$
$$\underset{S}{\diagdown\diagup} \qquad\qquad \longrightarrow \qquad\qquad (6)$$
$$+ \; CH_2\!=\!CH\text{-}CH_3$$

$$\sim\!\!\sim\!CH_2SNa \; + \; CH_2\text{-}CH\text{-}CH_3 \longrightarrow \sim\!\!\sim\!CH_2\text{-}S\text{-}CH_2\text{-}\overset{CH_3}{\underset{|}{CH}}\text{-}SNa$$
$$\underset{S}{\diagdown\diagup} \qquad\qquad\qquad (7)$$

---

*Monomer in THF is added by flow instead of by distillation. (Ed.)

Table 4.— Block Copolymers with three blocks : Butadiene-Propylene Sulfide, Isoprene-Propylene Sulfide and 2 Vinyl Pyridine-Propylene Sulfide

Initiator	1st. Monomer added	2nd. Monomer added	Molecular weights for the homopolymer			Molecular weights for the copolymer			
			$\overline{M}_n$ th.	$\overline{M}_n$ osm.	$\overline{M}_w$	$\overline{M}_n$ osm.*	$\overline{M}_n$ calc.	$\overline{M}_n$ osm.	$\overline{M}_w$
sodium naphtalene	butadiene	propylene sulfide	72000	80000	99000	152000	169000	170000	200000
sodium biphenyl	isoprene	propylene sulfide	100000	110000	133000	193500	234000	252000	260000
sodium biphenyl	2 vinyl pyridine	propylene sulfide	78000	85000	105000	151000	169000	170000	202000

*Molecular weight measurements on unpurified copolymer. (Ed.)

CONCLUSION

The carbanions of polyisoprene and polybutadiene are slowly deactivated in polar solvents, but this destruction may be sufficiently slow at low temperature to permit the preparation of block copolymers. Two types of active species may be seen, the initial one changing relatively rapidly (and faster at higher temperature) into the second "isomerized" form. However, both forms initiate the polymerization of their own monomer or other monomers. The stability of the living ends of poly-2 vinylpyridine and polypropylene sulfide are also very good, so that it is possible to synthetize well defined block copolymers having narrow molecular weights distribution with the following couples of monomers : isoprene - butadiene, isoprene - propylene sulfide, butadiene - propylene sulfide, 2 vinylpyridine - propylene sulfide.

Many copolymers can be now synthetized (30) and are interesting materials for various studies. The phenomenon of incompatibility of blocks leads to important results in dilute solutions (31) and gives organized structures in gel (32, 33) and also in bulk (34).

REFERENCES

1 - M. SZWARC, Nature, 178, 1168 (1956)
2 - M. SZWARC, M. LEVY and R. MILKOVITCH ,
                J. Am. Chem. Soc., 78, 2656 (1956)
3 - M. SZWARC, Carbanions Living Polymers and Electron
                Transfer Process, Interscience, New-
                York, chapt. 5
4 - P.J. FLORY, Principles of polymer chemistry,
                Cornell University Press, Ithaca,
                New York, chapt. 8
5 - S. BYWATER, A.F. JOHNSON and D.J. WORSFOLD ,
                Can. J. Chem., 42, 1255 (1964)
6 - A. GOURDENNE and P. SIGWALT ,
                European Polymer J., 3, 481 (1967)
7 - G. SPACH, M. LEVY and M. SZWARC ,
                J. Chem. Soc., 355 (1962)
8 - A. GOURDENNE and P. SIGWALT,
                to be published
9 - M. FONTANILLE and P. SIGWALT,
                European Polymer J., 5, 553 (1969)

10 - A. REMBAUM, SHIAO-PING SIAO and N. INDICTOR,
            J. Polymer Sci., 56, 517 (1962)
11 - S.N. KHANNA, M. LEVY and M. SZWARC,
            Trans. Faraday Soc., 68, 747 (1962)
12 - A. GOURDENNE,
            (unpublished results)
13 - G. CHAMPETIER, M. FONTANILLE and P. SIGWALT,
            International Symposium on Macromole-
            cular Chemistry, IUPAC, Tokyo-Kyoto
            N°IV, 55 (1966)
14 - A.A. KOROTKOV, L.A. SHIBAEV, L.M. PYRKOV,
     V.G. ALDOSHIM and S.I. FRENKEL,
            Vysokomol. Soed., 1, 443 (1959)
15 - S. SCHLICK and M. LEVY,
            J. Phys. Chem., 64, 883 (1960)
16 - R.J. ANGELO, R.M. IKEDA and M.L. WALLACH,
            Polymer, 6, 141 (1965)
17 - P. SIGWALT,
            Bull. Soc. Chim., 3, 423 (1964)
18 - W. BUSHUK and H. BENOIT,
            Can J. Chem., 36, 1616 (1958)
19 - A. GOURDENNE and P. SIGWALT,
            Bull. Soc. Chim., 10, 3678 (1967)
20 - M. MORTON and L.J. FETTERS,
            J. Polymer Sci., A2, 3311 (1964)
21 - R.S. STEARNS and L.E. FORMAN,
            J. Polymer Sci., 41, 381 (1959)
22 - M. MORTON, E.E. BOSTICK, R.A. LIVIGNI and
     L.J. FETTERS,
            J. Polymer Sci., A 1, 1735 (1963)
23 - D.J. WORSFOLD and S. BYWATER,
            Can. J. Chem., 42, 2884 (1964)
24 - A. REMBAUM, F.R. ELLS, R.C. MORROW and
     A.V. TOBOLSKY,
            J. Polymer Sci., 61, 155 (1962)
25 - S. BOILEAU, G. CHAMPETIER and P. SIGWALT,
            Makromolekulare Chem., 68, 180 (1963)
26 - P. SIGWALT,
            Main lecture, International Symposium
            on macromolecular Chemistry, IUPAC,
            Budapest (1969)
27 - A. GOURDENNE and P. SIGWALT,
            Bull. Soc. Chim., 10, 3685 (1967)
28 - S. BOILEAU and P. SIGWALT,
            Compt. Rend., 261, 132 (1965)
29 - M. FONTANILLE, A. GOURDENNE and P. SIGWALT,
            to be published.
30 - L.J. FETTERS, Polymer Symposia, 26, 1 (1969)

31 - H. BENOIT, Robert A. Welch Symposium on Polymers
            (1966)
32 - A. SKOULIOS and G. FINAZ,
            Compt. Rend., <u>252</u>, 3467 (1961)
33 - C. SADRON,
            Angew. Chem., <u>75</u>, 472 (1963)
34 - A.J. KOVACS and B. LOTZ,
            Kolloid-Z.Z. Polymere <u>209</u>, 97 (1966)

CONTROLLED BLOCK COPOLYMERIZATION OF TETRAHYDROFURAN AND NEW

ASPECTS OF TETRAHYDROFURAN POLYMERIZATION KINETICS

T. Saegusa, S. Matsumoto, and Y. Hashimoto

Department of Synthetic Chemistry, Faculty of

Engineering, Kyoto University, Kyoto, Japan

Summary

A new method (phenoxyl end-capping method) is presented for
the determination of the concentration of propagating species [p*]
in the cationic polymerization of tetrahydrofuran (THF), in which
the cationic propagating end was quantitatively converted into the
corresponding phenyl ether by treatment with sodium phenoxide.
The amount of phenyl ether group was determined by UV spectroscopy.
The examination of the [p*] change during the THF polymerization by
various catalyst systems showed that the course of polymerization
is very much dependent on the nature of the catalyst. Some catalysts induce polymerization without termination, whereas some others
cause decrease of [p*] by termination. From time-[p*] and time-
conversion data, the rate constant of propagation kp was deter-
mined according to $\ln[([M]t_1 - [M]e)/([M]t_2 - [M]e)] = kp \int_{t_1}^{t_2}[p*]dt$,
where [M]t's and [M]e are the instantaneous and equilibrium
monomer concentrations, respectively, and the integrated value of
[p*] was obtained by graphical integration. An important conclusion was that the kp value was affected very little by the nature
of the catalyst. Big difference in the pattern of the [p*] change
during the polymerization is ascribed mainly to the differences in
the rates of initiation and termination. On the basis of the
results of kinetic studies, the block copolymerization of THF with
3,3-bis(chloromethyl)oxetane (BCMO) was carried out, in which the
second polymerization was initiated by the living growing end of
poly-THF prepared by using BF_3-ECH as living polymerization catalyst. Two types of block copolymers were prepared; one is A-B
type block copolymer consisting of poly-THF (A) and poly BCMO (B)
blocks, and the other is A-B-A' type block copolymer in which two
crystalline blocks of poly-THF (A) and poly BCMO (A') were bound

to each end of the central rubbery block of the THF–BCMO random copolymer (B). By the application of the phenoxyl end-capping method, the lengths of each blocks in copolymers were calculated from the [p*] value. The A–B–A' type block copolymer, whose structure resembles that of the so-called "thermolastic" polymer, exhibited properties characteristic of a vulcanized rubber.

INTRODUCTION

The present paper is concerned with the cationic ring-opening polymerization of tetrahydrofuran (THF). The first topic is the kinetic study on the basis of the determination of real concentration of the propagating species, [p*], and the second topic is the application of the [p*] determination to the block copolymerization of THF with 3,3-bis(chloromethyl)oxetane (BCMO). Since the first report of H. Meerwein et. al. in 1937,[1] a large number of catalysts have been found for the THF polymerization. The apparent activity of these catalysts as judged from the reaction rate and the molecular weight of polymer product depends very much on the nature of the catalyst. Some catalysts produce very high molecular weight products in high yields, whereas some ones give only low molecular weight polymers. Recently, several kinetic studies on the THF polymerization have been reported.[2-5] These studies, however, were justified only in living polymerization systems which did not suffer from chain transfer and termination. In the present study, a new method for the [p*] determination was developed, which was successfully applied to the kinetic analyses of both living and non-living polymerization systems with several catalysts. An important finding is that the rate of propagation is not affected by the nature of catalyst.

The living polymerization system of THF was rightly selected on the basis of kinetics by the above method, which was utilized to prepare the block copolymer of THF with BCMO, i.e., the living propagating end of the THF polymerization was subjected to the initiation of BCMO polymerization. The lengths of the THF and BCMO blocks were calculated by knowing [p*]. An A–B–A' type block copolymer was also prepared, in which two crystalline blocks of poly-THF and poly-BCMO were bound to each end of the central rubbery block of THF–BCMO random copolymer. This A–B–A' type block copolymer exhibits properties of vulcanized rubber.

NEW ASPECTS OF POLYMERIZATION KINETICS

The effect of the nature of catalyst systems on the rate of propagation reaction was investigated by means of the [p*] determination. The propagation reaction of the THF polymerization is a reversible process.

$$\text{(1)}$$

The rate is given by the equation,

$$-d[M]/dt = k_p[p^*][M] - k_{-p}[p^*] \tag{2}$$

where $[M]$ and $[p^*]$ are the concentrations of monomer and propagating species, respectively, and k_p is the propagation rate constant. The rate constant of depolymerization, k_{-p}, in eq. (2) is related to k_p by means of the equilibrium monomer concentration, $[M]e$, as follows.
At equilibrium,

$$k_p[p^*][M]e = k_{-p}[p^*] \tag{3}$$

therefore,

$$k_{-p} = k_p[M]e \tag{4}$$

By using eq. (4), eq. (2) now becomes

$$-d[M]/dt = k_p[p^*] \left\{ [M] - [M]e \right\} \tag{5}$$

The value of $[M]e$ at $0°$ has been determined experimentally to be 1.7 mole/l[3].

In the kinetic analysis of the propagation reaction, the problem is how to determine $[p^*]$ value. Sometimes, the catalyst concentration itself was taken as $[p^*]$. As will be described later, this is not valid in most cases. In the $[p^*]$ determination, the polymerization system is treated with sodium phenoxide at a desired time of reaction and the cationic propagating end is converted quantitatively into the corresponding phenyl ether. The concentration of phenyl ether group thus formed at the polymer end is determined by UV spectroscopy.

$$\text{(6)}$$

By this method, the instantaneous concentration of propagating species is known.

New Method of $[p^*]$ Determination by Phenoxyl End-Capping[6]

For the phenoxyl end-capping method according to eq. (6), following three requisites should be satisfied.
  1. The molar extinction coefficient of the phenyl ether group at polymer end should be known.
  2. The conversion of the propagating chain end into phenyl ether should be quantitative and instantaneous.
  3. Any side reaction producing phenyl ether group should not occur.

Extinction coefficient. The molar extinction coefficient (εmax) of the phenyl ether group at polymer end was estimated from the εmax of two phenyl alkyl ethers, phenetole and ω-methoxybutyl phenyl ether.

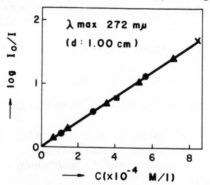

In methylene dichloride and in diethyl ether, the two phenyl ethers showed the same absorption spectra of λmax at 272 mμ with the same ε max. The Lambert-Beer plots for the two phenyl ethers fell on a single straight line as is shown in Fig. 1. Methoxybutyl phenyl ether is a model compound of the phenoxyl end-capped polymer of THF having a degree of polymerization of one. It is clear that the presence of ether linkage in the alkyl group does not affect the εmax of a phenyl alkyl ether. In addition, the UV spectrum of the phenoxyl end-capped poly-THF strongly resembles the spectra of these two phenyl ethers. From these findings, the εmax of these phenyl ethers, 1.93×10^3 1/mol.cm is reasonably assingned to the phenyl ether group at polymer end.

Quantitative conversion. The extent of the reaction of cyclic oxonium ion at the propagating chain end with sodium phenoxide was examined in a model reaction using triethyloxonium tetrafluoroborate (Et_3OBF_4).

The oxonium salt in ethylene dichloride was treated with excess sodium phenoxide at room temperature for 1 hr in the same way as that for the [p*] determination and the amount of phenetole was determined by UV spectroscopy. The results in TABLE I show that the reaction 7 is almost quantitative. Therefore it is reasonable to assume that the cyclic oxonium ion at the propagating end is quantitatively converted into the corresponding phenyl ether.

Figure 1. Extinction coefficient of phenyl alkyl ethers: $O(CH_2)_4OCH_3$: ($\blacktriangle$) in CH_2Cl_2 and ($\bullet$) in Et_2O; OEt : ($\times$) in CH_2Cl_2 and ($\blacksquare$) in Et_2O.

TABLE I

Reaction of Sodium Phenoxide with Et_3OBF_4 at Room Temperature

⟨O⟩-ONa[1]	Et_3OBF_4[2]	⟨O⟩-OEt		Δ ⟨O⟩-ONa[4]
$\times 10^{-5}$ mol	$\times 10^{-5}$ mol	$\times 10^{-5}$ mol	Yield (%)[3]	
5.64	1.85	1.83	99	1.63
5.64	2.18	2.14	98	1.96

[1] In 5 ml of THF.   [2] In 5 ml of $(CH_2Cl)_2$.   [3] Based on Et_3OBF_4.
[4] Amount of sodium phenoxide consumed.

In a separate experiment, it has also been established that the reaction 7 is complete in less than 20 sec at 0°. Cyclic trialkyloxonium ion is more strained and its reactivity toward phenoxide ion is even higher than that of the triethyloxonium ion. Therefore, it is reasonably assumed that the propagating species is converted immediately into phenyl ether.

Side reactions. A series of reference experiments were made to confirm the absence of possible side reactions. When epichlorohydrin (ECH) was treated with sodium phenoxide under the same conditions, no phenyl ether was formed and sodium phenoxide was not consumed. In the second experiments, the polymerization systems with various catalysts were first terminated by sodium hydroxide and then treated with sodium phenoxide. In these cases, formation of phenyl ether was negligible. Thus the three requisites for the [p*] analysis have been satisfied.

The method of [p*] determination by phenoxyl end-capping was then applied to the THF polymerization by $AlEt_3$-H_2O (2:1)-ECH system. The [p*] values were determined at several polymerization times by the phenyl ether concentration at polymer end and by the consumption of sodium phenoxide during the phenoxyl end-capping reaction. The results are shown in TABLE II. There is a good agreement between [p*] values from phenyl ether concentration and from the consumption of sodium phenoxide.

In our previous paper,[7] the THF polymerization by $AlEt_3$-H_2O-ECH system was shown to be a living polymerization. Each polymer molecule has a propagating center at the end and [p*] corresponds to the molar concentration of polymer, which can be calculated from the amount of polymer and its molecular weight according to the equation.

$$[p^*] = \frac{\text{(gram of polymer)}}{\text{(mol. wt.)} \times \text{(volume of system)}} \qquad (8)$$

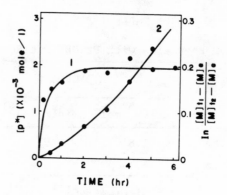

Figure 2. Polymerization of THF by $AlEt_3$-H_2O (2:1)-ECH:
(1) time-[p*] curve; (2) first-order plot: conditions
are same as that in TABLE II.

TABLE II

[p*] Values in the THF Polymerization by $AlEt_3$-H_2O (2:1)-ECH.[1]

Polymzn time hr	Polymer yield %	10^3 [p*] (mol/l.)		
		from ⬡-OR[2]	from △⬡-ONa[3]	from $\bar{M}_n$[4]
1/6	—	1.23	—	—
1/2	1.0	1.48	1.40	—
1	2.5	1.63	1.52	—
2	5.6	1.87	1.80	—
3	8.7	1.83	1.60	—
4	13.2	2.15	2.02	2.39
5	18.3	1.96	2.03	2.15
6	21.7	1.98	2.03	2.04

[1] Bulk polymerization at 0°; [M]o, 12.6 mol/l.; [$AlEt_3$]o, 0.18
mol/l.; [ECH]o, 0.02 mol/l.   [2] Amount of phenyl ether group.   [3] Amount of sodium phenoxide
consumed.   [4] Molecular weight determined by osmometry.

The [p*] values given by eq. (8) were in good agreement with those
obtained by the phenoxyl end-capping method.   These results are
taken as justifying the phenoxyl end-capping method.

Fig. 2 shows clearly that [p*] increases gradually in the

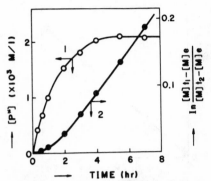

Figure 3. Polymerization of THF by BF$_3$-ECH:(1) time-[p*] curve; (2) first order plot:bulk polymerization at 0°;[M]o,12.6 mol/l.; [BF$_3$]o,0.011 mol/l.;[ECH]o, 0.010 mol/l.;[M]e,1.7 mol/l.[3]; t$_1$=0.

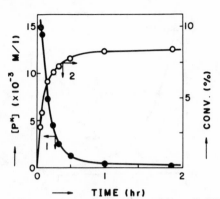

Figure 4. Polymerization of THF by SnCl$_4$-ECH:(1) time-[p*] plot; (2) time-conversion curve:bulk polymerization at 0°;[M]o,12.6 mol/l.;[SnCl$_4$]o,0.055 mol/l.; [ECH]o,0.053 mol/l.

induction period (first stage) and then remains unchanged throughout the subsequent stage of polymerization (second stage). The induction period shown by the time-[p*] relationship corresponds to that by the first order plot.

Polymerization by Lewis Acid-ECH[8]

Kinetic analysis of the THF polymerization by the binary catalyst system of a Lewis acid and ECH was carried out using the phenoxyl end-capping method. As the Lewis acid component, boron fluoride (BF$_3$), stannic chloride (SnCl$_4$) and ethylaluminum dichloride (AlEtCl$_2$) were employed. Polymerizations were carried out at 0° and mostly in bulk.

Fig. 3 shows the time-[p*] curve and the first order plot of the polymerization by BF$_3$-ECH. [p*] is increased slowly in the early stage and then it remains almost constant in the later stage. Polymerization by this catalyst system involves a slow initiation and proceeds without termination. The first order plot in Fig. 3 shows a gradual increase of polymerization rate to a constant value, which corresponds to the time-[p*] relationship. It should be noted that the constant value of [p*] is about 20 % of the initial catalyst concentration.

The course of the THF polymerization by SnCl$_4$-ECH is shown in Fig. 4. The time-[p*] curve indicates a rapid decrease of [p*], which corresponds to the rapid decrease of conversion rate. These data clearly demonstrate the occurrence of rapid termination to stop the polymerization at a very low conversion. Correspondingly, the polymer produced in this polymerization was of low molecular weight.

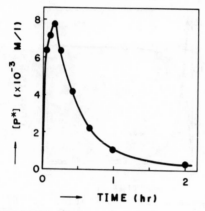

Figure 5. Polymerization of THF
by $AlEtCl_2$-ECH :time-[p*] curve:
bulk polymerization at $0°$; [M]o,
12.6 mol/1.; $[AlEtCl_2]$o, 0.054 mol/1.,
[ECH]o, 0.051 mol/1.

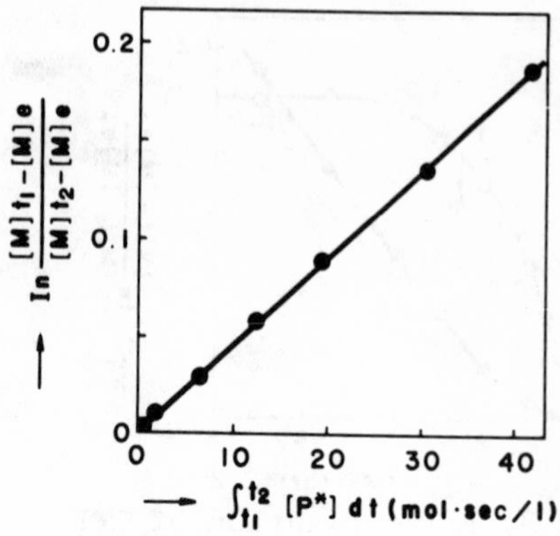

Figure 6. Polymerization of
THF by BF_3-ECH: plot of eq. 9:
conditions are given in Fig. 3.

Fig. 5 shows the time—[p*] curve of the polymerization by
$AlEtCl_2$-ECH, which is similar to that by $SnCl_4$-ECH. In this case,
however, a period for the increase of [p*] is observed in the early
stage of polymerization. This indicates that the initiation is a
little slower. The polymer produced by this catalyst system was
also of low molecular weight.

Now, the kp values of these polymerizations were determined
in the following way. Integration of the rate equation 5 with
respect to time gives

$$\ln \frac{[M]t_1 - [M]e}{[M]t_2 - [M]e} = kp \int_{t_1}^{t_2} [p*]dt \qquad (9)$$

where $[M]t_1$ and $[M]t_2$ are the monomer concentrations at time t_1
and t_2, respectively. The cumulative value of [p*] in eq. 9 can
be obtained by graphical integration of the time—[p*] curve.
The kp value is to be obtained from the linear plot of eq. 9.

Fig. 6 shows the linear relationship of eq. 9 (t_1=0,
t_2=variable) for the THF polymerization by BF_3-ECH. From the slope
of this straight line, the kp value was calculated (TABLE III).
As shown in Figs 7 and 8, the linear relationship of eq. 9
(t_1=3 min) was observed even in the polymerizations by $SnCl_4$-ECH
and by $AlEtCl_2$-ECH, where a drastic decrease of [p*] took place
(Figs 4 and 5).
TABLE III summarizes the kp values of the polymerizations by
the three Lewis acid-ECH systems together with the values for other
catalyst systems. As has been shown previously, these three catalyst systems differ from one another in the time—[p*] relationship.

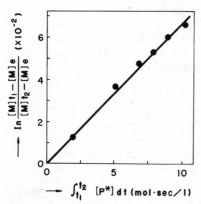

Figure 7. Polymerization of THF by SnCl$_4$–ECH : plot of eq. 9: conditions are given in Fig. 4; [M]e, 1.7 mol/1.[3]; t$_1$ = 3 min.

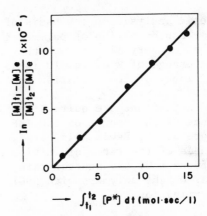

Figure 8. Polymerization of THF by AlEtCl$_2$–ECH : plot of eq. 9 : conditions are given in Fig. 5; [M]e, 1.7 mol/1.[3]; t$_1$ = 3 min.

TABLE III

Propagation Rate Constants of THF Polymerization[1]

Catalyst	$10^3 \cdot$ kp (1./mol·sec)
BF$_3$[2]	4.3
SnCl$_4$[2]	6.2
AlEtCl$_2$[2]	7.8
AlEt$_3$[2]	5.6
AlEt$_3$–H$_2$O (2:1)[2]	10
Et$_3$OBF$_4$	3.6

[1] Polymerization at 0°.   [2] In combination with ECH.

However, their kp values are quite similar to each other. In addition, these kp values are quite close to the kp value for the AlEt$_3$–H$_2$O–ECH system.[9] In the propagation reaction of eq. 1, the counter anion, A$^\ominus$, is derived from the Lewis acid component of initiator. Therefore it is reasonably concluded that the rate of propagation is affected very little by the nature of counter anion. This conclusion is similar to the observation in the THF polymerization by the AlEt$_3$–H$_2$O–ECH system. The apparent activity of this system expressed by the $\underline{k}$ value of the following equation changes very much according to the H$_2$O/AlEt$_3$ ratio.

$$-d[M]/dt = k[AlEt_3]o \left\{ [M] - [M]e \right\} \tag{10}$$

The kinetic analysis on the basis of eq. 5 by knowing [p*] from the molar concentration of polymer gave the conclusion that the kp value changes very little with the change of the $H_2O/AlEt_3$ ratio. Thus the change of k of eq. 10 is ascribed mainly to the change of the efficiency of the formation of the propagating species.

These findings are quite significant to manifest that the big difference of the course of the THF polymerization according to the nature of the Lewis acid catalyst is ascribed mainly to the differences of the rates of initiation and termination but not to the difference of propagation rate. Thus the formation of low molecular weight polymers with $SnCl_4$ and $AlEtCl_2$ is attributed to rapid termination.

Polymerization by Trialkyloxonium Salt

Trialkyloxonium salt is also a powerful initiator for the THF polymerization. Kinetics of the polymerization initiated by triethyloxonium tetrafluoroborate (Et_3OBF_4) was analyzed by the phenoxyl end-capping method. In this case, however, phenetole is formed from the remaining initiator, if it remains unreacted in the system, by the process of phenyl ether end-capping.

$$(11)$$

Therefore, phenetole should be separated from the polymer phenyl ether before [p*] determination. After several trials, a method of separation was successfully found, in which phenetole was removed by vacuum distillation together with decalin as a distillation entrainer. By means of this method, the concentrations of the two cationic species, the triethyloxonium initiator and the propagating species, were determined separately.

Fig. 9 shows the curves of time — [p*] and time — initiator concentration relations of the polymerization at 0°. From these relationships, two important findings can be pointed out. Firstly, the polymerization by Et_3OBF_4 is not a living polymerization. The time — [p*] curve shows the occurrence of a slow termination. Secondly, the initiator is consumed fairly rapidly.

According to the same procedure as used for Lewis acid-ECH systems, kp was calculated from the linear relationship of eq. 9

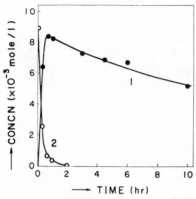

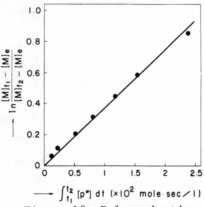

Figure 9. Polymerization of THF by
Et_3OBF_4: (1) time-[p*] curve;
(2) decrease of the concentration
of Et_3OBF_4: polymerization at 0°
in methylene dichloride;[M]o,
6.3 mol/l.;$[Et_3OBF_4]$o,9.0 ×10^{-3}
mol/l.

Figure 10. Polymerization of
THF by Et_3OBF_4: plot of eq.9:
conditions are given in Fig. 9;
[M]e,1.7 mol/l.; t_1 =0.

shown in Fig. 10. The kp value is included in TABLE III. Here
again, the kp value for Et_3OBF_4 is very close to the values for
other catalyst systems.

CONTROLLED BLOCK COPOLYMERIZATION OF THF WITH BCMO

Synthesis of Block Copolymer

On the basis of the detailed kinetic studies by [p*] determination, the controlled block copolymerization of THF with BCMO
was explored. A living polymer of THF was prepared by using living
polymerization catalyst and the second polymerization was initiated
by the cationic living end of poly THF. For this purpose, BF_3-ECH
system was employed, because this system has been confirmed by
the present study to induce the living polymerization of THF.
In addition, BF_3 was suitable also from our previous finding[10] that
the BF_3OEt_2 catalyzed polymerization of THF-BCMO mixture gave copolymers in which the THF and BCMO units were randomly distributed.

A problem here was that the THF monomer is not consumed completely because of the equilibrium in the propagation step.
In order to prepare a copolymer having poly-BCMO block, therefore,
THF must be removed from the reaction system without causing the
depolymerization of the living poly-THF produced. For this purpose,
THF was removed by vacuum distillation at lower temperatures in
the presence of BCMO. The depolymerization of THF was quenched
sufficiently at lower temperatures.

In this way, two types of block copolymers were prepared.
The one consists of two blocks, poly-THF block and poly-BCMO block.

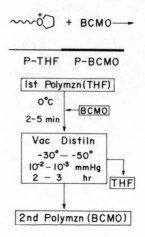

Figure 11. Scheme of two-stage block copolymerization.

The other is composed of three blocks, poly-THF block, a block of
THF-BCMO random copolymer and poly-BCMO block. Fig. 11 shows
schematically the procedure for the preparation of the first block
copolymer of A-B type. The THF polymerization was carried out at
0° for a designed period of time. The length of the THF block was
controlled according to the results of the kinetic studies.
BCMO was added to this system and the mixture was cooled quickly
to -50° to quench the reaction. Then the remaining THF was removed
in vacuo and the system was warmed to -30°, where the second stage
polymerization of BCMO was performed. At the end of the first stage,
[p*] was determined by means of the phenoxyl end-capping method.
    An example of two stage block copolymerization was as follows.

    1st stage

THF	37 mmol (3 ml)	
n-heptane	7 ml	
BF$_3$:THF	0.2 mmol	0°, 18 hr
ECH	0.1 mmol	

    p*, 3.3×10^{-2} mmol; yield of polymer, 0.63 g
    THF removal at -30°~-50° for 2.5 hr

  2nd stage

BCMO	5 mmol (0.6 ml)

      -30°, 3 hr
      yield of polymer 1.02 g

In the process of THF removal, the solvent was also distilled out of the reaction system and the second stage polymerization proceeded apparently in solid state.

The product of two stage block copolymerization was resinous and it was soluble in hot chloroform and insoluble in ethanol. Since the BCMO homopolymer is insoluble in hot chloroform and poly-THF is soluble in ethanol, the solubility character of this product is taken to show the formation of a copolymer.

The presence of the poly-THF and poly-BCMO blocks in the copolymer was clearly shown by IR and DSC analyses of the product. In Fig. 12, the IR spectrum of the two-stage copolymerization product is compared with the spectra of poly-THF and poly-BCMO and with that of a random copolymer. The spectrum of the block copolymer possesses the crystalline bands of poly-BCMO[11] at 860 cm^{-1} and 890 cm^{-1} and the crystalline band of poly-THF at 1000 cm^{-1}. Furthermore, the spectrum of the two-stage copolymerization product has no absorption band at about 870 cm^{-1}, which is characteristic for the THF-BCMO random copolymer. In the other region of the spectrum, the two-stage copolymer was different from the random copolymer. DSC (Differencial Scanning Calorimetry) diagram of the block copolymer is shown in Fig. 13, which demonstrates two endothermic peaks at 29° and 169° corresponding to the crystalline melting temperatures of poly-THF block and poly-BCMO block, respectively. These data clearly shows that the two-stage copolymer is a block copolymer composed of poly-THF and poly-BCMO blocks.

On the basis of the p* value, the lengths of two blocks were calculated. The block copolymer of the above example was characterized as follows; poly-THF block: DP, 260 (MW, 19000); poly-BCMO block: DP, 75 (MW, 12000). This corresponds to 22 mole % of BCMO content. In this case, however, chain transfer between

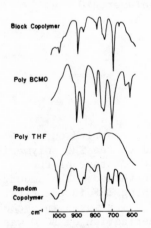

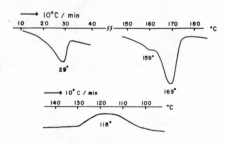

Figure 12. Comparison of IR spectra of homopolymers and copolymers of THF and BCMO (film).

Figure 13. DSC diagram of the block copolymer, (poly-THF)-(poly-BCMO) : heating rate, 10°/min.

the growing chain end and the ether linkage in the polymer molecule occurs to a small extent. Therefore, these values are approximate ones.

The second block copolymer is A-B-A' type one, in which the two ends of the internal amorphous block of the random copolymer (B) are bound to the two crystalline blocks of poly-THF (A) and poly-BCMO (A'). This structure corresponds to that of the so-called "thermolastic" polymer.

The procedure of the preparation of this block copolymer is similar to that of the first one. In this case, however, random copolymerization of THF and BCMO was allowed to proceed for a designed period of time before the removal of THF at low temperature. The [p*] determination was made at the ends of the first and the second stages to calculate the lengths of the blocks. A typical example is shown below.

1st stage (THF polymerization)

THF	185 mmol (15 ml)	
methylcyclohexane	35 ml	0°, 13 hr
BF_3:THF	1.0 mmol	
ECH	0.6 mmol	

        p*, 0.20 mmol; yield of polymer, 3.3 g

2nd stage (random copolymerization)

THF	140 mmol	0°, 1.5 hr
BCMO	92 mmol (11 ml)	

        p*, 0.13 mmol (65%); yield of polymer, 10.6 g
        THF removal at $-30° \sim -50°$ for 3 hr.

3rd stage (BCMO polymerization)

        0°, 2 hr
        yield of polymer 18.3 g

The first and second stage polymerizations were carried out in methylcyclohexane as solvent. BCMO was added to the THF polymerization mixture of the first stage and the second stage of the THF/BCMO copolymerization was subsequently performed. The amount of THF in the second stage was that of remaining THF in the first stage. After reaction for 1.5 hr at 0°, THF was removed in vacuo at low temperature and the third stage polymerization of BCMO was carried out at 0°. In this way, 18.3 g of block copolymer was obtained. The product of the three stage block copolymerization was elastic material which was soluble in hot chloroform or in hot THF and insoluble in ethanol. In Fig. 14, the IR spectrum of

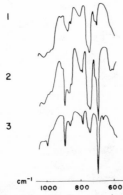

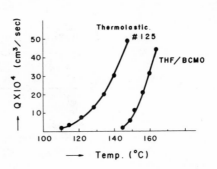

Figure 14. IR spectra of block copolymers having different structures
(film):(1) (poly-THF)-(random copolymer);(2) (poly-THF)-(random copolymer)-(poly-BCMO);(3) (poly-THF)-
(poly-BCMO).

Figure 15. Thermoplasticity
of block copolymers:
heating rate,3°/min;
Q, flow rate.

the product is compared with the spectra of other two types of
block copolymers. The top one consists of poly-THF block and
the block of THF-BCMO random copolymer, and the bottom one is
the two stage block copolymer having poly-THF and poly-BCMO blocks.
In the spectrum of the A-B-A' type block copolymer, there are
crystalline bands of poly-BCMO at 890 cm^{-1} and 700 cm^{-1} as well as
the band at 870 cm^{-1} characteristic of the random copolymer.
The above data as well as the mechanical properties of the copolymer described below strongly indicates that this copolymer has
a structure of A-B-A' type.
      From the p* values of the first and the second stages, the
lengths of the each blocks were calculated. The values obtained
for the product of the above experiment are given in TABLE IV.
Unfortunately, the block copolymerization suffers from termination
in the second stage and a part of the product is the block copolymer
which consisted of only poly-THF block and random copolymer block.
Hence, the lengths of the second and third blocks are to be taken
as the averages. Exploration for a better catalyst systems which
does not cause termination in the second and the third stages as
well as the problem of chain transfer to polymer molecule will
be the subject of future investigations.

Mechanical Properties of A-B-A' Type Block Copolymer

      Some mechanical properties of the A-B-A' type block copolymer
were examined. The measurements were made on the copolymer sample
shown in TABLE IV. In Fig. 15, the thermoplasticity of the THF-
BCMO block copolymer of A-B-A' type is compared with that of Shell's
"Thermoplastic", which is a A-B-A type block copolymer based on
polybutadiene (block B) and polystyrene (block A).

TABLE IV

Block Structure of A-B-A' Type Copolymer

	1st block	2nd block	3rd block
structure	poly-THF	random copolymer	poly-BCMO
Mol. Wt.	1.7×10^4	$\sim 4 \times 10^4$	6.2×10^4
DP	240	—	400
p* (mmol)	0.2	$0.13^{1)}$	
1-2-3, 65% ; 1-2, 35%			

1) p* at the end of the second stage.

Here, Q represents the rate of flow. The thermoplasticizing temperature of the THF-BCMO block copolymer is about 20° higher than that of the butadiene-styrene block copolymer. This data shows that the copolymer is not chemically cross-linked.

Fig. 16 shows the stress-strain behavior at room temperature. Here, the THF-BCMO block copolymer is compared with the butadiene-styrene block copolymer, the vulcanized and unvulcanized SBR.

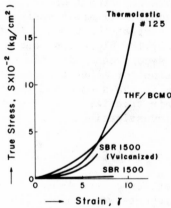

Figure 16. Stress-strain curves of various rubbers : at room temperature; speed, 50 cm/min.

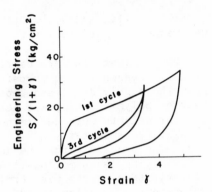

Figure 17. Hysteresis curves of the A-B-A' type
THF-BCMO block copolymer in the first and the third
cycle of stretching : at room temperature.

It is seen that both block copolymers behave like a vulcanized
rubber. However, the THF-BCMO block copolymer shows a large hysteresis loss, particularly in the first cycle of stretching, as
shown in Fig. 17. In addition, a yield point is observed in the
first cycle. As the cycle is repeated, the hysteresis loss becomes
considerably small and the yield point disappears. This behavior
seems to be the characteristics of a physically cross-linked block
copolymer. In the third cycle, the behavior of the THF-BCMO block
copolymer is similar to that of vulcanized SBR. The relaxation
spectrum also showed that the THF-BCMO block copolymer behaves like
a vulcanized rubber.

All these data are taken as supporting the A-B-A' type structure of the THF-BCMO block copolymer, i.e., (poly-THF)-(THF/BCMO
random copolymer)-(poly-BCMO).

EXPERIMENTAL SECTION

Materials. Monomers and solvents were commercial reagents
which were purified in the following ways. The methods of purification of THF, ECH and methylene dichloride have been described
before.[7,8] BCMO was purified and dried by repeated distillation
over calcium hydride under reduced nitrogen pressure. n-Heptane
and methylcyclohexane were purified by the usual method and
distilled repeatedly over sodium-potassium alloy. Commercial
reagents of $SnCl_4$ and $AlEtCl_2$ were purified by distillation under
reduced nitrogen pressure. $AlEt_3$-H_2O system,[7] BF_3-THF complex[12]
and Et_3OBF_4[13] were prepared and purified according to the reported
procedure. The sodium phenoxide solution was prepared by treating

a THF solution of phenol with metallic sodium.[6] Other reagents
were purified by the usual methods. Materials were stored under
a nitrogen atmosphere.

Kinetic experiments. Polymerizations were carried out under
an atmosphere of dry nitrogen. Reagents were transferred by means
of a syringe which had been thoroughly dried and kept in a nitrogen-
filled container. The rate of polymerization was determined gravi-
metrically. Osmotic molecular weights were determined by means
of a Mechrolab Model 502 membrane osmometer.

Determination of [p*]. The procedure for the determination
of [p*] by the phenoxyl end-capping is as follows. At a desired
time of reaction, the polymerization system was terminated by the
addition of excess sodium phenoxide solution in THF and the mixture
was allowed to react for an appropriate period of time. In the case
of the THF polymerization, the reaction mixture was then treated
with an aqueous solution of sodium hydroxide and the polymer was
extracted with CH_2Cl_2. After drying by potassium carbonate, the
CH_2Cl_2 extract was subjected to UV measurement. In the [p*]
determination of THF-BCMO copolymerization (second stage in the
A-B-A' block copolymerization), the short-stopping mixture was
poured into a large amount of methanolic sodium hydroxide solution.
The precipitated polymer was filtered, washed with methanol, re-
dissolved in CH_2Cl_2 and subjected to UV measurement. Reference
experiments showed that polymer phenyl ether is not lost during
these procedures. UV spectra were taken with a Shimadzu SV-50A
spectrophotometer.

THF-BCMO block copolymerization. Reactions were carried out
in a flask directly attached to a vacuum system with magnetic
stirring. The first stage polymerization of THF was performed
as usual in solution in n-heptane or methylcyclohexane under
nitrogen atmosphere. When n-heptane was used as solvent, the system
became heterogeneous owing to precipitation of poly-THF, which was
again dissolved on addition of BCMO. BCMO was added to this mixture
by means of a syringe. In the two-stage copolymerization, the
mixture was cooled in a Dry Ice-methanol bath to -50° immediately
after the BCMO addition. In the three-stage copolymerization,
the THF-BCMO random copolymerization was carried out at 0° after
the BCMO addition, and then the mixture was cooled to quench the
reaction. The removal of the THF monomer was continued at -50°
until the pressure of the system was below 10^{-2} mmHg (usually 2-3
hr). Then the third stage polymerization of BCMO was carried out
in vacuo without introducing nitrogen. Block copolymers were
isolated by precipitation with methanol and, when necessary,
purified by reprecipitation using a solvent-precipitant pair of
chloroform-ethanol. IR spectra of the copolymers were taken at
room temperature in a state of film. DSC measurements were done
by a Perkin-Elmer DSC-1B calorimeter with a heating rate of 10°/min.

Mechanical properties. Mechanical properties of the A-B-A'
type block copolymer were measured on a copolymer sample containing
0.3% of phenyl-β-naphthylamine as an antioxidant. The total

molecular weight of the sample was about 1.2×10^5. The thermo-plasticity was measured by a Shimadzu Koka flow tester with a constant heating rate of 3°C/min. For stress-strain measurements, ring-shaped specimens were cut from 1 mm sheets that had been molded at 150°C. Measurements were made at room temperature using Shimadzu Autograph model IM-500 with a crosshead speed of 50 cm/min. Hysteresis curves were obtained similarly.

The authors are indebted to Dr. S. Kusamizu and his group of Japan Synthetic Rubber Co., for the measurements of mechanical properties of the block copolymer.

References

1. H. Meerwein and E. Kroning, J. prakt. Chem., [2], 147, 257 (1937).
2. B. A. Rozenberg, E. B. Ludvig, A. R. Gantmakher, and S. S. Medvedev, J. Polym. Sci., Part C, No. 16, 1917 (1967).
3. E. A. Ofstead, Polym. Preprints, 6, 674 (1965).
4. D. Vofsi and A. V. Tobolsky, J. Polym. Sci., Part A, 3, 3261 (1965).
5. D. Sims, Makromol. Chem., 98, 235; 245 (1966).
6. T. Saegusa and S. Matsumoto, J. Polym. Sci., Part A-1, 6, 1559 (1968).
7. H. Imai, T. Saegusa, S. Matsumoto, T. Tadasa, and J. Furukawa, Makromol. Chem., 102, 222 (1967).
8. T. Saegusa and S. Matsumoto, Macromolecules, 1, 442 (1968).
9. T. Saegusa, H. Imai, and S. Matsumoto, J. Polym. Sci., Part A-1, 6, 459 (1968).
10. T. Saegusa, T. Ueshima, H. Imai, and J. Furukawa, Makromol. Chem., 79, 221 (1964).
11. G. Wasai, T. Saegusa, J. Furukawa, and H. Imai, Kogyo Kagaku Zasshi, 67, 1428 (1964).
12. R. C. Osthoff, C. A. Brown, and J. A. Hawkins, J. Amer. Chem. Soc., 73, 5480 (1951).
13. H. Meerwein, E. Battenberg, H. Gold, E. Pfeil, and G. Willfang, J. prakt. Chem., 154, 83 (1939).

POLYMERIZATION OF CYCLIC IMINO ETHERS. VIII. BLOCK COPOLYMERS*

M. Litt**
Department of Macromolecular Science
Case Western Reserve University
Cleveland, Ohio 44106

J. Herz**
Eastern Research Laboratories
Stauffer Chemical Corporation
Dobbs Ferry, New York

and

Edith Turi
CCRL
Allied Chemical Corporation
Morristown, New Jersey 07960

INTRODUCTION

It was recently reported that poly(N-acyl ethyleneimines) when crystalline, have surfaces whose critical surface tension, γ_c, ranges from 24.5 dynes/cm. for a butyryl side chain to 22 dynes/cm. for hexanoyl and longer side chains (1). Thus, the polymer surface must be composed of close packed methyl groups. It was also found recently that poly(N-acyl trimethyleneimines) where the acyl group was relatively short have relatively high conductivities at 50% relative humidity. In static electricity decay tests, the charge half-life was between 5 and 0.5 seconds depending on the side chain (2). Since the polymerization is non-terminating (3) one could easily make block copolymers and investigate the manner in

* This paper appeared in the Symposium on Block Copolymers as, "Block Copolymers of Cyclic Imino Ethers", by M. Litt and J. Herz.
** This research was performed at Allied Chemical Corporation, where the authors were employed at that time. We thank Allied Chemical for permission to present the work.

which the different components organize themselves. We therefore
chose two polymers whose physical properties contrasted violently,
as the candidates for block copolymers. These were poly(N-lauroyl
ethyleneimine) [M.P.=150°C, γ_c = 22 dynes/cm., water insoluble (1,3)]
and poly (N-acetyl trimethyleneimine) [amorphous, water soluble,
Tg = 30°C, static decay half life ≈ 0.5 sec. (2,4)]. The
polymerization reaction is given below.

† T_sOCH_3 = Methyl p-toluene sulfonate

Chain transfer to monomer occurs in the alkyloxazoline system (2)
to the extent of about one to every three hundred polymerizations.
In order that all the molecules should be true blocks, the degree of
polymerization was kept very low for the first section of the block.
The second section may have some chain transfer, producing a small
amount of homopolymer in addition to the blocks.

EXPERIMENTAL

Synthesis of the block copolymers

This example is representative of the whole series of block
copolymers synthesized. The exact polymerization conditions are
found in Table I.

Into a carefully cleaned and dried three neck round bottom

flask equipped with a stirrer, reflux condenser and an addition
funnel was added 12.5 g. (0.0556 moles) of undecyloxazoline and
0.80 g. (0.00430 moles) of methyl p-toluenesulfonate. The
reaction mixture was heated and stirred in an atmosphere of
nitrogen at 125° for 35 minutes. 37.0 g. of 2-methyloxazine*
(0.373 moles) was then added from the addition funnel. Heating and
stirring were continued for an additional 2.5 hours. The total
monomer to catalyst ratio is 100.

TABLE I

Synthesis Conditions for Block Copolymers

	M/I^a		Wt%		Polymerization Conditions		
	Component		Component		Time (min)		Temp. °C
Copolymer	A^b	B^b	A	B	A	B	
1	5.9	39.4	25	75	13	60	125
2	6.3	14.1	50	50	20	100	115
3	6.2	4.6	75	25	14	90	125
4	12.9	87	25	75	35	145	125
5	12.1	25.5	50	50	35	85	125
6	13.0	9.8	75	25	35	85	125

a) M/I is the ratio of moles of monomer to moles of initiator.
With the above polymerization conditions it is equivalent to the
degree of polymerization.

b) Component A is undecyl oxazoline [— poly(N-lauroyl ethyleneimine)]. Component B is methyl oxazine [— poly(N-acetyl
trimethyleneimine)].

Static Decay Testing

The block copolymers were dissolved in chloroform (about 1%
solution), and a taffeta cloth woven from Nylon 6 fiber was dipped
in the solution. After evaporation, the cloth was conditioned at
50% relative humidity for at least one day before testing.
Measurements were made at 73°F and 50% R.H. The cloth was attached
to a Rothschild static voltmeter (type R-1019) and was charged
by means of a high voltage source within the meter to 100V. The
time taken to dissipate half the initial charge was recorded.

---

*In this paper, 2-methyloxazine refers to 2-methyl-4,5-dihydro-
oxazine. (Ed.)

Critical Surface Tensions of Block Copolymers

Thin films of copolymer were deposited on cleaned glass slides from 1% chloroform solutions. Contact angles of droplets of a number of aromatic and aliphatic liquids with the polymer were measured using the sessile drop technique of Shafrin and Zissman (5).

Melting Points and Heats of Fusion

The copolymers were annealled and the melting points of the crystalline phase was measured by D.T.A. using a duPont 900 Differential Thermal Analyser. More recently heats of fusion of annealled homopolymer were measured in a Perkin Elmer Differential Scanning Calorimeter, Model 1B. The sample was heated to 160°C for one minute, cooled to 130°C and annealled for varying lengths of time under N_2. It was then cooled at 10°C/min to room temperature and heated at 20°C/min. to measure heat of fusion.

RESULTS AND DISCUSSION

We follow the view that the morphology of a block copolymer is determined by the phase that organizes first. If this is a crystalline phase, lamellae tend to form with the amorphous portion on the surface of lamellae (6). In the case presented here, the crystalline phase crystallizes under conditions where the amorphous phase can still flow due to solvent inclusion. Thus we expect the lamellae to be reasonably highly organized with the amorphous phase surrounding them. The size and perfection of the lamellae should depend on the degree of polymerization and the time of annealling of the crystalline polymer.

The critical surface tensions (γ_c) of the annealled block copolymers in every case was 22 dynes/cm. showed that the surface in all cases was that of the crystalline, low γ_c phase. This might be expected from the above discussion. Thus the film can arrange itself to have the lowest possible energy content, which implies low surface energy.

This view is borne out by the static electricity half-lives (Table II).

TABLE II

Static Decay Half-Life Times For Block Copolymers at 50% R.H.

Copolymer	Time (sec.) 100V $\rightarrow$ 50V
1	10±0
2	9±0
3	31±0
4	7±0
5	160±14
6	>1,800
Uncoated fabric	>1,800

The decay half-life, of pure poly(N-acetyl trimethyleneimine) is about 0.5 seconds. The lowest value for the block copolymers is 7 seconds. Thus, the pathways for electrical conduction are either highly tortuous or have interruptions. The degree of organization of the crystalline regions is markedly changed on increase of the degree of polymerization from six to 13. At the low value, there is reasonable though poor continuity of the conducting phase at all copolymer compositions. The organization of the crystalline phase is obviously patchy, though good enough to produce the low critical surface tension. At the higher degree of polymerization, 13, there is reasonable conductivity at 75 weight % of poly(N-acetyl trimethyleneimine). At lower proportions, the conducting portions of the block copolymer are isolated from each other almost completely, showing that the crystalline phase is the matrix.

The polymer melting points can clarify the picture still further. As can be seen in Table III these are related only to the degree of polymerization and not at all to the weight fraction of crystalline polymer.

TABLE III

Melting Points of Poly(N-Lauroyl ethyleneimine) as a Block Polymer

Component	Crystalline Block	
Copolymer	M/I $(=P_n)$	T_m °C
1	5.9	79
2	6.3	86
3	6.2	86
4	12.9	120
5	12.1	115
6	13.0	120
Homopolymer	4,550	150

Thus initial crystallization is probably in lamellar form as was found with other crystalline/amorphous block copolymers (6). However, the perfection and size of the lamellae seem to depend on the degree of polymerization. At the high degree of polymerization, if the crystalline phase is 50% or more of the block copolymer, it can grow so as to almost completely isolate portions of the amorphous fraction. At the low degree of polymerization, large lamellae apparently do not form under the conditions of these experiments, and the amorphous phase is only partly isolated even when it is only 25 weight % of the total. Thus there is some conductivity.

There is one further interesting point, not connected with the block nature of the copolymers. From the melting points versus the degree of polymerization, one can calculate ΔHu and ΔSu of melting, following Flory (7). A graph of this plot is shown in Figure 1.

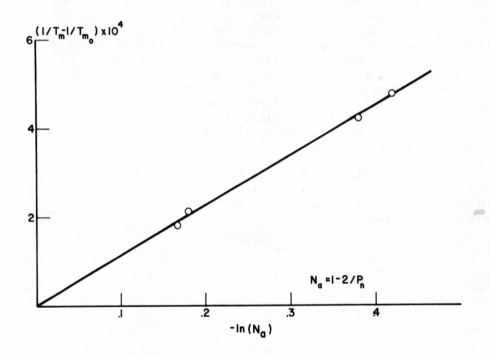

FIG I. MELTING POINTS VERSUS DEGREE OF POLYMERIZATION FOLLOWING FLORY (7). $1/T_m - 1/T_{m_0} = (R/\Delta Hu) \ln N_a$. $N_a = 1 - 2/P_n$

We find ΔHu = 1,800 cal/mole., while ΔSu = 4.24 cal/mol. degree C. This is extremely low for a polymer whose repeat unit includes 14 carbon atoms, and indicates a high degree of order in the melt of the crystalline phase.

In order to confirm this very low heat of fusion, heats of fusion were measured on annealled homopolymer by DSC. Unannealled chips, on melting absorbed 1,220 cal/mole. After annealling for one hour at 130°C, a heat of fusion of 1,760 cal./mole was measured, and after fifteen hours this had risen to 1,820 cal./mole. As we know from x-ray photographs that the polymers show almost no amorphous scattering after annealling, we feel that the value of ∿1,800 cal./mole for the heat of fusion of poly(N-lauroyl ethyleneimine) is a valid one. Two different methods of measurement confirm the value.

References

1. M. Litt and J. Herz, J. Colloid and Interface Science - in press.
2. Unpublished results.
3. T. G. Bassiri, A. Levy and M. Litt, Polymer Letters 5, 871 (1967).
4. A. Levy and M. Litt, Polymer Letters 5, 881 (1967).
5. E. G. Shafrin and W. A. Zisman, J. Colloid Sci., 7, 166 (1952).
6. C. Sadron., Angew. Chem., 75, 472 (1963) and later references by A. Skoulios, A. Kovacs, and A. Keller.
7. P. J. Flory, Principles of Polymer Chemistry, pp. 568-576, Cornell University Press., N.Y., 1953.

THERMAL STABILITY OF PIPERAZINE BLOCK COPOLYMERS

Stephen D. Bruck*
National Heart and Lung Institute
National Institutes of Health
Bethesda, Maryland 20014

Ashok Thadani
Chemical Engineering Department
The Catholic University of America
Washington, D. C. 20017

Previous publications discussed the thermal stability of a series of piperazine polyamides (1-4). In the case of the piperazine homopolymers, it was shown that methyl substitution of the piperazine rings effectively decreases their thermal stability in a vacuum. On the other hand, the presence of terephthaloyl moieties in the chain increases the thermal stability of the system in comparison with the polymers that have isophthaloyl moieties in their chains. An essentially random thermal breakdown was indicated from the nature of the volatile degradation products, the character of the rate curves, and the high activation energies.

This paper deals with the thermal properties of various piperazine block copolymers having number average molecular weights in the range of 13,000 to 30,000.

Experimental Details

Materials

Block copolymers of terephthaloyl trans-2,5-dimethylpiperazine/isophthaloyl trans-2,5-dimethylpiperazine, terephthaloyl trans-2,5-dimethylpiperazine/sebacyl trans-2,5-dimethylpiperazine, and isophthaloyl trans-2,5-dimethylpiperazine/sebacyl trans-2,5-dimethylpiperazine were prepared by a method described by Morgan and Kwolek (5).

---

*Inquires concerning this paper should be addressed to this author.

Viscosity Measurements

The inherent viscosities, which were determined in m-cresol at 30± 0.05°C with Cannon-Fenske viscometers, varied between 1.2 and 2.2. These values correspond to a number average molecular weight of approximately 14,000-30,000 (3).

Thermogravimetry

Programmed and isothermal degradation studies were carried out with samples of 3-7 mg using a Cahn RG electrobalance (1-4) with a modified hang-down tube-thermocouple assembly.

Volatile Degradation Products

The volatile degradation products were analyzed mass spectrometrically by a fractionation method already described (1,2).

Results and Discussion

The following abbreviations will be used in this paper:

t-2,5-DiMePip-T/t-2, 5-DiMePip-I denotes the block copolymer of terephthaloyl trans-2, 5-dimethylpiperazine/isophthaloyl trans-2, 5-dimethylpiperazine;

t-2,5-DiMePip-T/t-2,5-DiMePip-10 denotes the block copolymer of terephthaloyl trans-2,5-dimethylpiperazine/sebacyl trans-2,5-dimethylpiperazine;

t-2,5-DiMePip-I/t-2,5-DiMePip-10 denotes the block copolymer of isophthaloyl trans-2,5-dimethylpiperazine/sebacyl trans-2,5-dimethylpiperazine.

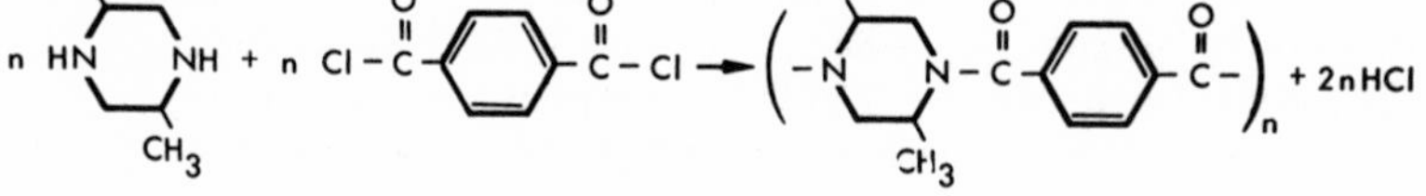

The block copolymers are prepared from the substituted piperazine and the appropriate diacid chloride and are composed of units of: terephthaloyl trans-2,5-dimethylpiperazine, isophthaloyl trans-2,5-dimethylpiperazine,

sebacyl trans-2,5-dimethylpiperazine,

The melting temperatures of the block copolymers are above 400°C as indicated by thermograms obtained by differential scanning calorimetry (Perkin-Elmer Model DSC-1B) and DTA measurements (DuPont Thermograph Model 920).

Thermal Degradation Under Programmed Heating Rates

The thermal degradation of the following block copolymers with number average molecular weights 13,000-14,000 were investigated under programmed rates of heating of 4°C/minute both in air and in a vacuum:

(a) t-2,5-DiMePip-T/t-2,5-DiMePip-I, 50/50 mole %;
(b) t-2,5-DiMePip-T/t-2,5-DiMePip-10, 50/50 mole %;
(c) t-2,5-DiMePip-I/t-2,5-DiMePip-10, 50/50 mole %.

The results are illustrated by Figure 1 (air) and Figure 2 (vacuum).

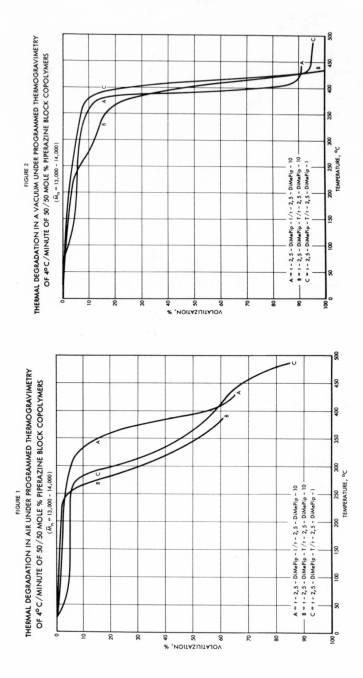

FIGURE 1

THERMAL DEGRADATION IN AIR UNDER PROGRAMMED THERMOGRAVIMETRY
OF 4°C/MINUTE OF 50/50 MOLE % PIPERAZINE BLOCK COPOLYMERS
($\bar{M}_n$ = 13,000 - 14,000)

A = t - 2,5 - DiMePip - I/t - 2,5 - DiMePip - 10
B = t - 2,5 - DiMePip - T/t - 2,5 - DiMePip - 10
C = t - 2,5 - DiMePip - T/t - 2,5 - DiMePip - 1

FIGURE 2

THERMAL DEGRADATION IN A VACUUM UNDER PROGRAMMED THERMOGRAVIMETRY
OF 4°C/MINUTE OF 50/50 MOLE % PIPERAZINE BLOCK COPOLYMERS
($\bar{M}_n$ = 13,000 - 14,000)

A = t - 2,5 - DiMePip - I/t - 2,5 - DiMePip - 10
B = t - 2,5 - DiMePip - T/t - 2,5 - DiMePip - 10
C = t - 2,5 - DiMePip - T/t - 2,5 - DiMePip - 1

It can be seen that the relative thermal stability in air of these
medium molecular weight block copolymers is:

> t-2,5-DiMePip-I/t-2,5-DiMePip-10 > t-2,5-DiMePip-T/t-2,5-
> DiMePip-I > t-2,5-DiMePip-T/t-2,5-DiMePip-10

while in a vacuum it is:

> t-2,5-DiMePip-T/t-2,5-DiMePip-I > t-2,5-DiMePip-I/t-2,5-
> DiMePip-10 > t-2,5-DiMePip-T/t-2,5-DiMePip-10.

Isothermal Degradation in Vacuum

The rates of thermal degradation were obtained from volatilization vs. time curves with the aid of an electronic computer.
The number average molecular weights of the block copolymers
varied between 13,000-30,000 and could be divided into two groups:
(1) high molecular weight samples ($Mn = 20,000-30,000$), and
(2) medium molecular weight samples ($Mn = 13,000-14,000$).
Figures 3 and 4 show, respectively, the rates of volatilization
of high molecular weight samples of 70/30 and 50/50 mole % block
copolymer of t-2,5-DiMePip-T/2,5-DiMePip-I (from reference 4).

FIGURE 4

RATES OF THERMAL DEGRADATION OF 50/50 MOLE %
BLOCK COPOLYMER ($\bar{M}_n$ = ABOVE 20,000)
OF t-2,5-DiMePip-T/t-2,5-DiMePip-I

from S.D. Bruck and A.A. Levi, J. Macromol. Sci.-Chem., A1 (6), 1095, (1967)

FIGURE 3

RATES OF THERMAL DEGRADATION OF 70/30 MOLE %
BLOCK COPOLYMER ($\bar{M}_n$ = ABOVE 20,000)
OF t-2,5-DiMePip-T/t-2,5-DiMePip-J

from S.D. Bruck and A.A. Levi, J. Macromol. Sci.-Chem., A1 (6), 1095, (1967)

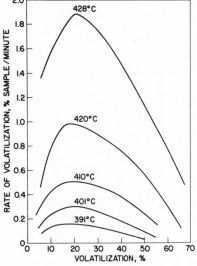

Figures 5 to 7 illustrate the rates of volatilization of <u>medium</u>
molecular weight samples of 50/50 mole % block copolymers of
t-2,5-DiMePip-T/t-2,5-DiMePip-I, t-2,5-DiMePip-T/t-2,5-DiMePip-10
and t-2,5-DiMePip-I/t-2,5-DiMePip-10, respectively.

FIGURE 5

RATES OF THERMAL DEGRADATION OF 50/50 MOLE % BLOCK COPOLYMER
OF t - 2, 5 - DiMePip - T/t - 2, 5 - DiMePip - I IN A VACUUM
($\bar{M}_n \approx 13,000 - 14,000$)

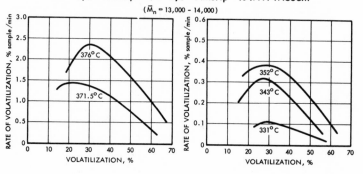

FIGURE 6

RATES OF THERMAL DEGRADATION OF 50/50 MOLE % BLOCK COPOLYMER
OF t - 2, 5 - DiMePip - T/t - 2, 5 - DiMePip - 10 IN A VACUUM
($\bar{M}_n \approx 13,000 - 14,000$)

FIGURE 7
RATES OF THERMAL DEGRADATION OF 50/50 MOLE % BLOCK COPOLYMER
OF t - 2,5 - DiMePip - I/t - 2,5 - DiMePip - 10 IN A VACUUM
$(\bar{M}_n = 13{,}000 - 14{,}000)$

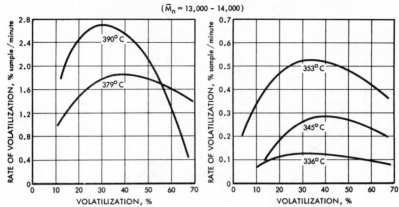

The data indicate that at comparable temperatures, the high
molecular weight samples show considerably lower rates of volatilization than the medium molecular weight specimen. This
situation is especially clear when the two 50/50 mole % composition block copolymers of t-2,5-DiMePip-T/t-2,5-DiMePip-I
are considered. At the temperature of 410°C, the high molecular
weight sample exhibits a maximum rate of volatilization of 0.51%
sample per minute, whereas at 412°C the medium molecular weight
material shows a maximum rate of 4.06% sample per minute. It
is apparent that the stability of the block copolymers is influenced by their molecular weights in contrast to the homopolymers discussed in earlier publications (1-3). The results also
indicate that with the exception of the medium molecular weight
50/50 mole % block copolymer of t-2,5-DiMePip-I/t-2,5-DiMePip-
10, the maxima of the rate curves appear between 25 and 30%
conversion. On the other hand, the maxima in the case of the
medium molecular weight 50/50 mole % t-2,5-DiMePip-I/t-2,5-
DiMePip-10 appear between 30 and 40% conversion. The significance of this will be discussed below.

     The Arrhenius plots for the thermal degradation of the
block copolymers in vacuum are illustrated by Figure 8, and
the pertinent data are summarized in Table 1.

FIGURE 8

ARRHENIUS PLOTS FOR THE THERMAL DEGRADATION IN VACUUM OF HIGH (20,000 - 30,000)
AND MEDIUM (13,000 - 14,000) $\bar{M}_n$ PIPERAZINE BLOCK COPOLYAMIDES

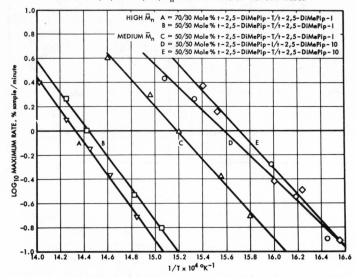

TABLE 1

RATES, ACTIVATION ENERGIES AND HALF-LIFE (T_h) VALUES FOR HIGH (20,000 - 30,000)
AND MEDIUM (13,000 - 14,000) $\bar{M}_n$ BLOCK COPOLYMERS OF t - 2,5 - DiMePip

POLYMER	TEMP °C	MAX RATE % sample/min	ΔE Kcal/mole	S sec^{-1}	T_h °C
T/I, 70/30 HIGH $\bar{M}_n$ *	440 429 419 411 401	2.50 1.22 0.76 0.42 0.20	61	1.9×10^{17}	436
T/I, 50/50 HIGH $\bar{M}_n$ *	428 420 410 401 391	1.88 0.90 0.51 0.29 0.16	60	1.5×10^{17}	427
T/I, 50/50 MEDIUM $\bar{M}_n$	412 395.5 385 370 360	4.06 1.96 0.98 0.42 0.20	51	4.7×10^{16}	393
T/10, 50/50 MEDIUM $\bar{M}_n$	376 371.5 352 343 331	2.37 1.44 0.38 0.32 0.12	49	7.2×10^{16}	370
I/10, 50/50 MEDIUM $\bar{M}_n$	390 379 353 345 336	2.70 1.86 0.53 0.29 0.13	44	8.1×10^{14}	373

*from S. D. Bruck and A. A. Levi, J. Macromol. Sci. - Chem., A1 (6), 1095, (1967)

The highest activation energies are exhibited by the high molecular weight samples of the block copolymers, while the lowest activation energy (44 Kcal/mole) is shown by the medium molecular weight 50/50 mole % t-2,5-DiMePip-I/t-2,5-DiMePip-10. This latter polymer also shows a shift in the maximum rates of volatilization to higher conversions, as mentioned above. In the case of purely random type degradations the maxima appear at about 26% conversion (6); the observed shift to 30-40% conversion coupled with a lowered activation energy indicates the influence of non-random processes, such as hydrolytic scission (7,8).

The half-life (T_h) values of the block copolymers correspond to 50% volatilization of the sample during isothermal conditions (9). According to the T_h values, the polymers can be arranged in the following order of relative thermal stability under isothermal conditions in a vacuum:

70/30 mole % high m.w. t-2,5-DiMePip-T/t-2,5-DiMePip-I $>$
50/50 mole % high m.w. t-2,5-DiMePip-T/t-2,5-DiMePip-I $>$
50/50 mole % medium m.w. t-2,5-DiMePip-T/t-2,5-DiMePip-I $>$
50/50 mole % medium m.w. t-2,5-DiMePip-I/t-2,5-DiMePip-10 $>$
50/50 mole % medium m.w. t-2,5-DiMePip-T/t-2,5-DiMePip-10.

Volatile Degradation Products

The results of the mass spectrometric analyses are summarized in Table 2.

TABLE 2

MASS SPECTROMETRIC ANALYSIS OF MAJOR DEGRADATION PRODUCTS VOLATILE AT
ROOM TEMPERATURE OF t-2,5-DiMePip BLOCK COPOLYMERS DURING VACUUM PYROLYSIS
($475°$C, 1 HOUR, ~10^{-5} mm Hg)

COMPONENT	70/30 T/I* ($\bar{M}_n$ = 20,000 to 30,000) mole %	50/50 T/I* ($\bar{M}_n$ = 20,000 to 30,000) mole %	50/50 T/I ($\bar{M}_n$ = 13,000 to 14,000) mole %	50/50 T/10 ($\bar{M}_n$ = 13,000 to 14,000) mole %	50/50 I/10 ($\bar{M}_n$ = 13,000 to 14,000) mole %
CARBON MONOXIDE	33.7	37.9	22.2	9.7	17.2
CARBON DIOXIDE	8.5	14.2	17.6	21.9	28.5
WATER	14.5	11.8	25.7	43.2 (?)	20.5
AMMONIA	16.4	14.0	1.6	0.2	0.1
NITROGEN	0.4	0.1	8.0	6.1	10.5
HYDROGEN CYANIDE	-	-	0.9	0.7	0.1
HYDROGEN	-	0.7	2.7	1.0	2.1
METHANE	13.3	8.1	10.8	2.9	8.2
Other Aliphatic Saturated and Unsaturated Hydrocarbons	5.5	4.2	5.5	9.6	8.5
2,5-DIMETHYL PYRAZINE	1.9	2.1	2.2	0.6	1.7
2-METHYL PYRAZINE	0.9	0.8	0.6	1.0	0.5
PYRAZINE	-	-	0.4	0.5	0.3
BENZENE	1.7	1.9	0.2	0.2	0.2

*from S.D. Bruck and A.A. Levi, J. Macromol. Sci.-Chem., A1 (6), 1095, (1967)

The main volatile degradation products consist of carbon monoxide, carbon dioxide, methane, hydrocarbons, ammonia and water.
The latter product most likely represents adsorbed moisture in
the polymer. The large quantities of carbon monoxide is unlike
the situation with aliphatic polyamides, where only trace amounts
of this degradation product are produced, while the appearance
of carbon dioxide has been attributed to hydrolytic processes
involving the amide groups (7-8). There is a greatly reduced
production of ammonia and increased evolution of nitrogen in the
case of the medium molecular weight polymers. It is interesting
to note that the medium molecular weight 50/50 mole % t-2,5-
DiMePip-I/t-2,5-DiMePip-10 showed that the largest quantity of carbon
dioxide (28.5 mole %) evolved during vacuum pyrolysis. Furthermore, this polymer also showed a shift in the maximum rates of
volatilization to higher conversions in comparison to the other
block copolymers, as discussed above. Such a shift is characteristic for polyamides in which non-random processes, such as
hydrolytic scissions, also operate and is accompanied by a lowering of the activation energy as well as by the appearance of
increased quantities of carbon dioxide. This view is supported
by the data of the present study inasmuch as the medium molecular weight 50/50 mole % t-2,5-DiMePip-I/t-2,5-DiMePip-10 block
copolymer shows the lowest activation energy (44 Kcal/mole), a
shift in the maximum rates of volatilization to higher conversion,
and the production of large amounts of carbon dioxide. Hence,
this polymer seems to be the most vulnerable to hydrolytic processes in comparison to the other samples studied.

Piperazine polyamides have other interesting properties
that might be useful in medical implant applications and these
are discussed in a related paper (10).

Acknowledgements

The mass spectrometric analyses were carried out by Mr.
Ernest Hughes and Mr. Julian Ives of the National Bureau of
Standards, Gaithersburg, Maryland. The computer work was done
by Mr. Stanley Favin, Applied Physics Laboratory, The Johns
Hopkins University, Silver Spring, Maryland.

This investigation was supported in part by Public Health
Service Research Grant No. 1-R01-GM-16192-01 from the Institute
of General Medical Sciences, National Institutes of Health, with
one of us (S.D.B.) as the Principal Investigator at The Catholic
University of America, Washington, D.C. Part of this paper
was presented at the Eighth Conference on Vacuum Microbalance
Techniques, Wakefield, Massachusetts, June 12-13, 1969.

References

1. S. D. Bruck, Polymer, (Lond.), 6, 483 (1965).
2. S. D. Bruck, Polymer, (Lond.), 7, 231 (1966).
3. S. D. Bruck, Polymer, (Lond.), 7, 321 (1966).
4. S. D. Bruck and A. A. Levi, J. Macromol. Sci.-Chem., A1 (6), 1095 (1967).
5. P. W. Morgan and S. L. Kwolek, J. Polymer Sci., A2, 181 (1964).
6. R. Simha and L. A. Wall, J. Phys. Chem., 56, 707 (1952).
7. S. Straus and L. A. Wall, J. Res. Natl. Bur. Std., 60, 39 (1958).
8. S. Straus and L. A. Wall, J. Res. Natl. Bur. Std., 63A, 269 (1959).
9. S. L. Madorsky, "Thermal Degradation of Organic Polymers," Interscience Publishers, New York, 1964.
10. S. D. Bruck, Polymer, (Lond.), in press.

INDEX